IEE POWER SERIES 18

Series Editors: Professor A. T. Johns
                J. R. Platts
                Dr. D. Aubrey

VACUUM SWITCHGEAR

VACUUM SWITCHGEAR

Allan Greenwood

The Institution of Electrical Engineers

Published by: The Institution of Electrical Engineers,
London, United Kingdom

The Institution of Electrical Engineers,
Michael Faraday House,
Six Hills Way, Stevenage,
Herts. SG1 2AY, United Kingdom

British Library Cataloguing in Publication Data

A CIP catalogue record for this book
is available from the British Library

ISBN 0 85296 855 8

Printed in England by Short Run Press Ltd., Exeter

Contents

Acknowledgments

A book is written by its author but it often happens that a vast number of people contribute indirectly. This was certainly the case with this book. Over the years I had countless discussions with physicists, metallurgists, fellow engineers, and yes, even managers. I must give special recognition to T. H. Lee with whom I burnt so much midnight oil on back porches in humid suburban Philadelphia. With three colleagues, G. Polinko, D. W. Crouch and H. R. Davidson, we started the interrupter development programme and learned together, often the hard way. Add to these P. Barkan, J. W. Porter, H. N. Schneider and W. R. Wilson and you have the group that got the project launched.

I would also like to recognise colleagues at the Research and Development Center, in particular G. A. Farrall, the late Gerhard Frind and L. P. Harris.

C. W. Kimblin (Westinghouse) and G. R. Mitchell gave me many insights through their publications.

Companies also have contributed; my special thanks go to General Electric which gave me the opportunity to break ground in this technology. I am indebted to many other companies – Calor Emag (ABB), Cooper Industries, Hitachi, Holec, Joslyn Hi-Voltage, Siemens, Toshiba and Westinghouse – which opened their factories to me.

Preface

The development of vacuum switchgear in the past 30 or 40 years has been unprecedented in the history of power switching technology. Vacuum switches existed in the 1950s, but they were very limited in their capability with respect to current carrying and current interruption. Today, in the early 1990s, vacuum dominates the medium-voltage sector for all power switching functions. In Japan, for example, vacuum commands almost 80% of the medium voltage switchgear market.

It was my good fortune to become intimately involved with the development of vacuum interrupters in the mid 1950s when I was part of a small team assigned to this pursuit at the General Electric Company. I have continued to be concerned with many aspects of vacuum switchgear ever since. This experience and this perspective have provided the basis for the production of this book. This should be evident from the first chapter, which is intended to be more than an historical record.

Chapters 2 to 4 delve into the science behind the technology and thereby provide the physical background for the remainder of the book. Vacuum arcs are different from gaseous arcs and current interruption in vacuum is different from interruption in gas circuit breakers. We need to understand these differences and the idiosyncracies of vacuum if we are to design, construct and use vacuum switchgear successfully.

I have striven to point out the learning process, the successes and failures along the way, and the occasional quantum jump, all of which have brought us to where we are today. I have endeavoured to make the treatment comprehensive so that the book will be useful to users, designers and manufacturers, alike. For the user there is an exhaustive chapter on applications which spans all devices from contactors through switches and reclosers to power circuit breakers. In addition, there is a much shorter chapter on maintenance. There are four chapters on different aspects of design and another on testing which should appeal to the designer. The chapter on manufacturing concentrates mainly on the interrupter because this is so entirely different from interrupters for oil and gas-blast breakers. The facilities required and the techniques used in their preparation and fabrication are much closer to those of the semiconductor industry. No material is included on traditional vacuum techniques since it is readily available elsewhere.

Allan Greenwood
Tortola, British Virgin Islands
January 1993

To Grace, my wife

Chapter 1

Historical review — how we arrived at where we are today

1.1 Introduction

The purpose of this chapter is to trace the development of vacuum switchgear from its beginnings to the very sophisticated state it has attained today (early 1990s). The enormous amount of work that has been done and the quite restricted space available here, prescribe a very limited account; in no way can it be comprehensive. What I hope to present instead are the major milestones, the successes and failures, and the important factors that have influenced the progress of the technology. The material will naturally be coloured, but hopefully not biased, by my own experiences in a long association with the subject.

The story starts around 1920, only 25 years after Thompson's discovery of the electron. Millikan and Shackelford had very recently discussed how electrons might be pulled from metals by strong electric fields. Coolidge had invented the X-ray tube so that reasonably good vacuum was possible within an enclosure, but sealing techniques were still problematical. The largely experimental and empirical work of Ayrton [1] represented much of the current knowledge of arcs; Compton [2] had not yet presented his treatment of the subject. Power switchgear technology was confined to oil and air-break devices.

One has the impression that the power switching equipment of the day was adequate, but that improvements would be necessary if technology was to keep pace with the growing demands of the utility industry. Unquestionably, these perceived needs were a strong factor promoting the study of vacuum as a switching ambient. We discover, in retrospect, that existing technologies had to continue to supply the industry's needs for many more years.

The suggested use of vacuum was met with some scepticism, but its proponents viewed it as the 'ideal medium' capable of producing the 'ideal switch'. This notion sprang from two beliefs: first, that vacuum provided the best-known dielectric, and secondly, that current interruption in vacuum would be a quite automatic process since vacuum provided nothing to support conduction. The first belief was no doubt based on X-ray tube experience, where a gap of the order of a centimetre could hold off over a 100 kV. This suggested that vacuum would be an excellent candidate for switching at high voltage, an opinion that still prevailed in the early 1950s when the author first became active in the field. 'Voltage would be easy', we said, 'interrupting a high current would be the problem'. We soon discovered, however, that post arc conditions of vacuum contacts are a far cry from the pristine state of polished X-ray tube electrodes and that the consequences for high voltage switches could be dire.

The second basic belief did not expect that the current would cease the moment the contacts parted. There would be a spark or arc, but many stated ideas regarding the nature of the arc in vacuum were very primitive. However, there was an early appreciation that gas on the surface and dissolved within the contacts could be released by arcing and thereby destroy the vacuum, and that means must be found to solve the problem. Parenthetically, it was this concept which prompted later investigators to use refractory materials, such as tungsten, molybdenum, and carbon, for contacts, believing that these could be heated to very high temperatures to expel the adsorbed and dissolved gasses. The material of the contacts and how it is processed is of paramount importance. The early choice of refractories for this purpose-made vacuum switches a reality, but undoubtedly retarded the development of vacuum circuit breakers. The distinction being made here relates to the magnitude of current being interrupted. A switch disconnects loads or isolates unloaded circuits, and as such usually handles at most a few hundred amperes. A circuit breaker is used to clear faults where the current can be many thousands or tens of thousands of amperes. For reasons discussed in due course, refractory contacts are incapable of this duty.

Another serious limitation for early investigators was instrumentation. Items that we now take for granted, or indeed, which we took for granted 30 years ago, simply did not exist. Electronics was in its infancy; the cathode ray oscilloscope was still quite primitive; means for measuring degree of vacuum were clumsy and limited in range. There was little in the way of equipment for residual gas analysis, and certainly devices such as scanning electron microscopes had not even been thought of. In the area of computation, the slide rule held sway.

One must conclude that conditions for launching an investigation into this new switching technology were not entirely auspicious.

1.2 Early pioneers

The first systematic investigation of vacuum as a switching ambient was carried out at the California Institute of Technology in the 1920s by Professor Royal Sorensen and his graduate students Mendenhall, Otis, Lindvall, Hamilton, and Haywood, with the essential support of their workshop staff[*]. In 1926 Sorensen and Mendenhall [3] reported on the results obtained from their first three switches at an AIEE meeting in Salt Lake City. What they had achieved was indeed impressive.

Figure 1.1 shows a line drawing of their second switch which was larger than the first but otherwise shared many of its characteristics. To open the switch, a bridging contact is caused to move vertically, creating two series breaks. Motion is initiated by the action of an external solenoid on an internal armature, thereby avoiding the need for a moving member to penetrate the vacuum envelope. Electrical connections to the fixed contacts entered through Housekeeper seals [5], the best glass-to-metal sealing technology available at the time. The switches were continuously pumped to maintain a vacuum of 10^{-6} torr.

---

[*] L. M. Burrage recently drew my attention to US Patent 441,542, dated November 25, 1890 in which O. A. Enholm describes a 'Device for transferring and controlling electric currents'. In this patent is the following '. . . I cause the primary circuit to be broken in a vacuum between solid conducting electrodes'. There is no evidence that the device was ever reduced to practice.

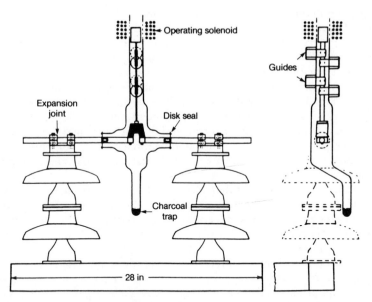

Figure 1.1   Instruction drawing for No. 2 Cal. Tech. vacuum switch [6]

(By permission of IEEE)

Switches 1 and 2 were subjected to a large number of switching operations which they performed extremely well. For example, switch 2 closed onto, and opened, 120 A in a 15 kV circuit 500 times without any sign of distress. Indeed, its performance was superior to an oil switch which was tested for a similar duty, in that interruption with the vacuum switch was always achieved at the first current zero, but not so the oil device.

Following these tests, switch 2 was sealed off from the vacuum system and left to sit for three months, after which it repeatedly interrupted a current of 600 A at 12.38 kV. A coasting synchronous condenser provided the source for these tests.

The complete reliance on gravity to close the switch contacts and maintain them closed, and the need to keep the solenoid energised to hold the switch in the open position, led to some design changes in the third prototype interrupter. This had a single, one inch break, but like the first two switches, the contacts were of copper. A copper bellows, or sylphon, permitted movement of one contact, a forerunner of the universal modern approach. Switch 3 was used to extend investigations to higher voltages. A current of 926 A was successfully interrupted at almost 41.6 kV.

The technological achievements of these three early switches were remarkable, but of equal importance was the scientific knowledge learned in the investigation. Insights were gained into the vacuum arc. For instance, it was found that the arc voltage was only 20 V or thereabouts, thus the energy dissipated in the arc was far less than in an oil switch. 'Conditioning' was observed, that is the progressive improvement in dielectric strength by spark breakdown of the contact gap, a process we now refer to as spark cleaning.

Work continued at Cal. Tech. through the remainder of the decade of the 1920s. An informative account of these activities is provided in a later review paper [4] and in Sorensen's discussion of Cobine's 1963 paper [6]. We learn that contacts of aluminum and tungsten as well as copper were used. Differences in performance were noted and predictions for other materials were made. For example, comments were offered on the gettering action of sputtered contact material as it accumulated as a thin film on the walls of the vessel. (The ability of such material to scavenge gas and thereby improve the vacuum.) Lindvall [7] concludes, 'Finally, the results obtained here suggest a definite line of future investigation, an intensive study of the inherent getter action of the vacuum switch when augmented by some sort of arcing contacts of metal more favorable than copper and by more active metallic wall deposits or adsorption surfaces, thus leading towards the ultimate goal, operation independent of pumping equipment'.

One of the most insightful statements made by these early pioneers appears in Hamilton's thesis [8] '. . . the theory based on the existence of a cathode spot, brought into being when the contacts open, and maintained by the action of the arc, explains the action of the switch in a satisfactory way and seemingly is in accord with all the observed phenomena'.

The work at Cal. Tech. continued for some time. Sorensen [4] describes a three-phase switch, with massive contacts - a blade engaging with spring-loaded fingers in the traditional style - which, not surprisingly, proved virtually impossible to degas. An interesting, unrelated comment was made regarding the puncture of the glass envelope and damage to bellows by focussed beams of electrons when a high voltage was maintained across the open gap of a switch.

1.3 Industrial follow-up and the doldrum years

In 1927 the General Electric Company was sufficiently encouraged by the progress made by Professor Sorensen and his associates to purchase patent rights and commence a program of its own under the direction of D.C. Prince. In due course, Haywood was hired to participate in this activity after he had completed his doctoral program at Cal. Tech.

According to Prince, many switches were built and tested and some successes were recorded. For example, one switch interrupted 3.31 kA at 48 kV, another cleared 5.28 kA at 14.5 kV. Cobine [6] reports that the most promising device was a continuously pumped metal-clad unit which ruptured 8.6 kA at 13.2 kV. This was perhaps the first device to use a radial magnetic field to produce arc rotation. However, the sealed interrupters were almost invariably plagued by gas evolution which accompanied arcing. Thus, after a number of operations, the vacuum deteriorated to the point where current interruption was no longer possible. There was also a realisation that practical interrupters would have to maintain their vacuum integrity over many years, whether operated or not, and at the time there was no instrument capable of detecting the small leaks that could destroy this integrity during a long shelf life. There were other problems: the propensity of clean copper contacts to become welded together; the generation of high overvoltages by the premature extinction of the arc between refractory contacts. More is said later about these phenomena in later chapters.

The difficulties just described, the general economic conditions in the early 1930s, and improvements in oil circuit breaker technology, caused the abandonment of the General Electric vacuum switching programme. Relatively little was published but much was learned, most notably the obstacles that barred the way to success.

There ensued a period of almost 20 years which can perhaps be best described as the doldrum years in that relatively little was done which directly concerned power switching in vacuum. This is not to say that there was no activity, at General Electric attention was turned to surge protectors with vacuum gaps, some of which were sold commercially. Yet these too languished, troubled by inconsistent breakdown characteristics. Kling [9] described what might best be called a vacuum relay for fast and repeated interruption of 10 A (AC or DC) at 250 V. In 1946, a thoughtful paper by Koller [10] reported on some work at the Berkeley campus of the University of California. He described experiments made with two switches, one sealed and one pumped, in which *direct currents* were interrupted. Breaking DC required reliance on the instability of the arc, but this was not a problem with the refractory contacts that were used (Mo, W, C) and the modest currents involved. Koller wrote of the development of a working pressure as the equilibrium condition between gas evolution by arcing and the capture of such gas by the metal vapour produced in the discharge. He identified pumping action by the vapour, that is, the capture of gas by the flux of diffusing metal atoms and the burying of the gas when the vapour reached and condensed on the contacts, shields, etc. Gettering by freshly deposited metal was also described. The electrode material used was quite gassy, so much was sputtered violently by exploding gas pockets. In many ways Koller's paper is perceptive, but the limited scope of his experiments led to some inaccurate predictions. For instance, he states, 'For larger currents, refractory metals are best suited'.

If relatively little was achieved directly with respect to vacuum switching technology during the doldrum years, a great deal occurred which had a quite crucial indirect affect. Techniques and processes were developed in other fields that in due course removed the barriers preventing the realisation of vacuum power switching devices. In this regard the doldrum years were an essential part of the evolution.

1.4 Emergence of the commercial vacuum switch

The 1950s witnessed an important development which had some far reaching consequences. I refer to the appearance in commercial production of vacuum switches which could be, and were, applied quite extensively in power systems. This event undoubtedly heightened the interest in vacuum as a power switching medium and contributed to its later wide acceptance as such. Moreover, it helped intensify efforts to extend vacuum switching capability into the power circuit breaker range.

The initiative for the development came from a new direction, namely the communications industry. Vacuum had been used there for some time in disconnecting devices, often in high frequency circuits. The current was usually very small but the voltage could be quite high. Antenna switches are a good example. It was natural to ask, 'Can the technique be extended to currents of interest to power system engineers?' Tests showed that indeed it could.

The 1956 paper of Jennings, *et al.* [11] reported some impressive achievements in capacitor bank switching. Besides single interrupter applications, series

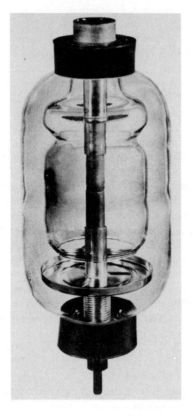

Figure 1.2   Early vacuum interrupter for load switching operation

(Courtesy Ross Engineering Corp.)

arrangements of two and four interrupters were described for operating voltages up to 230 kV. For the most part the current carried by the switch was limited to 200 A, which was, and remains, quite adequate for many installations. Other examples described had the vacuum interrupters bypassed by a metal blade, allowing a continuous current of 600 A without thermally overloading the vacuum device. When required to do so, the vacuum switch could interrupt the current once the bypass switch opened and commutated the current into its contacts. This technique is now widely used in modern switches.

The interrupters used at that time were made by the Jenning's Manufacturing Corporation, and were essentially adaptations of the existing communication switches. Like these devices, they had a glass envelope but a glass shield was added to avoid the deposition of metal vapour on the major insulation. Another metal shield or guard protected the bellows from incandescent particles thrown off from the contacts during arcing. The contacts themselves were of tungsten, 0.5 inch in diameter. The contact gap was 0.125 inch or thereabouts. Figure 1.2 illustrates what I am trying to describe.

Tungsten was chosen as a contact material because of the relative ease with which it could be degassed by heating to incandescence in vacuum. However, this procedure, which was performed by induction heating, was only partially successful inasmuch as gas was released when the magnitude or duration of the current was increased and this ultimately rendered the switches inoperative. For example, the switch depicted in Figure 1.2 had a maximum interrupting rating of 2000A.

A subsequent paper by Ross [12] in 1958 gave more information on the capacitance switching application and some instructive material was brought into the discussion by Schwager [13] who wrote of preinsertion resistors to reduce inrush current in bank-to-bank switching. Figure 1.3 shows an example of a capacitor switch with two interrupters in series, *circa* 1959. The smaller

Figure 1.3   1959 series application of vacuum interrupters for parallel switching of capacitor banks

(Courtesy of Joslyn Hi-Voltage Corp.)

cylindrical attachments are resistors that are inserted in closing to reduce the inrush current when switching one bank against another.

Ross also reported on fault current interruption, but the data was restricted to currents of a few thousand amperes. His approach to magnetising current switching was very slow opening over the first part of the stroke, with the intent of dissipating the systems inductive stored energy in a rapid series of reignitions and thereby avoiding overvoltages.

Several companies purchased Jennings interrupters and incorporated them in hardware for capacitance switching and other switching duties.

A photograph of the first vacuum device called a fault interrupter is shown in

Figure 1.4   Fault interrupting device rated 115 kV, 4 kA

(Courtesy of Joslyn Hi-Voltage Corp.)

Figure 1.4. It was designed for 115 kV and was solenoid operated. This was built in 1959 for New Orleans Public Service and was first tested at their Almonaster substation. The fault interrupt rating was 4000 A with 600 A continuous rating. The interrupter stack comprised five Jennings RL10G interrupters, which were 5 inches in diameter and were housed in 18 in modules.

At least one other manufacturer, Macklett Laboratories Inc., made and marketed vacuum interrupters which were applied in power systems by the Allis Chalmers Manufacturing Company [14].

This work of the 1950s was a triumph for engineering rather than science. It resulted in the manufacture and the marketing to a conservative utility industry of switching equipment that met important needs and complied with the industrial standards of the day. This state of affairs was brought about in spite of what we clearly recognise in retrospect as a very limited knowledge of the fundamentals of the vacuum arc. Subsequent publications showed that a niche had been established for vacuum switchgear.

1.5 The first vacuum power circuit breakers

In 1952 the General Electric Company initiated a new look at current interruption in vacuum. There was a belief at that time that the problems that had beset previous efforts could now be solved as a consequence of technological advances in other fields. Work began first in the corporate research and development laboratory in Schenectady, where focused research was directed to the composition and processing of contact materials, the better measurement of vacuum, especially pressure transients associated with vacuum arcs, and improvements in methods of leak detection. There were also experimental and theoretical studies of the vacuum arc itself. These concerned arc stability, electrode processes and the buildup of dielectric strength following arcing. For reasons of business security, much of this material was not reported at all outside of the company, or else it was published some years later [6].

In 1955 the Schenectady effort was substantially augmented by the formation of a second applied research group in the laboratory operation of the company's switchgear and control division in Philadelphia. To this group was assigned the task of designing, building and testing prototype vacuum interrupters for power circuit breaker applications. It was therefore more engineering-oriented. However, it also conducted its own theoretical and experimental research projects. It was the author's good fortune to be a part of the Philadelphia effort from its inception.

Still later, several engineers from the product department that would ultimately have responsibility for the design, engineering, manufacturing and marketing of any apparatus that might be produced, were secunded to the laboratory operation. They participated in all phases of the programme, and in due course returned to their former department to become an integral part of the operation that designed the first product and set up the facilities to produce it. This product was announced in 1962 [15], thus, ten years had elapsed since the vacuum program had been restarted.

The work in Philadelphia in the initial stages was a combination of three parallel efforts. The first was directed to the study of contact materials and contacts, the second to the design, construction and testing of vacuum interrupters, and the third to theoretical studies on vacuum arc phenomena and vacuum breakdown. The Schenectady programme had already shown that zone refined metal satisfied the need for essentially gas-free contact material. This procedure had been developed by the young semiconductor industry as a means of preparing super-pure silicon and germanium. The material is placed in a long boat-shaped crucible and then heated to melting by an induction coil. However, only a short section is molten at any time. Relative motion between the boat and the coil causes the molten zone to traverse the specimen from one end to the other, and as it does so, it sweeps impurities with it. This is a consequence of differential diffusion. After a number of passes, most of the impurity is isolated at the downstream end. This end can be removed and discarded. The pertinent fact in the present context is that dissolved gasses can be similarly swept to one end of the ingot, leaving behind material with a gas content of no more than a few parts per billion.

As discussed in Section 5.3.1 there are many other properties required of a contact material such as weak welds, good dielectric strength after arcing, low chopping level, etc. All these were addressed by the engineers and metallurgists in Philadelphia. Working with a demountable system, the author personally tested dozens of materials, pure metals, alloys, intermetallic compounds, sintered mixtures etc., for current chopping. He also tested a vast number of transformers in Pittsfield, MA, Rome, GA, and Fort Wayne, IN, for their response to current chops. In retrospect, an inordinate amount of time was spent on the current chopping problem, to the extent that it did not turn out in practice to be the serious shortcoming that we had supposed.

Designing, building and testing prototype interrupters, was a formidable challenge for people relatively inexperienced in vacuum techniques; we learned from experience, often the hard way. Each new device incorporated all recent knowledge that had been gained in the parallel fundamental experiments and analytical studies. In the beginning the cycle from start of design to completion of tests was about six months. This was shortened as time went on, and it became necessary to start on new designs before previous ones had been completely evaluated. It should also be pointed out that no interrupters could be made without first designing and constructing the tools, vacuum systems, bakeout ovens, degassing facilities etc., to do the job.

Special mention should be made of two patents which greatly enhanced interrupter performance and greatly influenced future interrupter design. The

Figure 1.5   Spiral contact for causing arc rotation [16]

first was the spiral contact [16], the second was the floating shield [17]. Figure 1.5 shows an example of a spiral contact, which are used in pairs, one being the mirror image of the other. The contacts mate on the annular rings, and the arc forms in this location when the contacts separate. When the arc is in the constricted mode, the current path is such that the self-generated magnetic field has a component that tends to drive the arc radially outward. When this happens, the arc roots move out along the 'petals', thereby enhancing the magnetic force just described. The arc is driven to the end of the petal, which curves around and is shaped to give a tangential component to the arc. As a result, the arc tends to jump from petal to petal, running around the periphery of the contact. This has two advantages; first, the arc never stays in any location for very long, and, second, the vapour tends to spread over a large surface when it condenses. As the technology expanded, much ingenuity was shown by competitors as they attempted to circumvent this patent.

A shield is necessary to protect the inside insulating walls of the vacuum interrupter from the arc products generated by current interruption. In high current interrupters the shield is metallic; this is clearly evident in Figure 1.6

Figure 1.6   Early prototype vacuum interrupter showing floating shield
(Courtesy of General Electric Co.)

which shows an early, hand-made prototype for the vacuum recloser [15], *circa* 1959. The fact that interruption is significantly improved by allowing the shield to float in potential, was a discovery made quite early in the course of the development programme and led to the aforementioned patent.

The principal accomplishments of the General Electric Company's vacuum interrupter development program during the 1950s are well reported in the paper of Lee, *et al.* [18], which the author had the satisfaction of presenting at the 1962 IEEE Winter Power Meeting in New York City. Only a few highlights are described here. Perhaps the most important milestone was the device shown in Figure 1.7, it was enclosed in SF_6 to avoid external flashover. The official designation of this switch was SS 5-1-1, but it will always be known to those

Figure 1.7   Experimental interrupter SS5-1-1, 'Sputnik'
(Courtesy of General Electric Co.)

involved in its development as 'Sputnik', because of its resemblance to the first Soviet space satellite which had recently been launched. It will be observed that the hemispherical shields form part of the vacuum envelope, and that electrical insulation is provided by glass bushings at both ends. The contacts were similar to that shown in Figure 1.5; the contact gap was 0.5 in. The interrupter was tested over the Christmas period in 1958 and the outcome was most encouraging. The results of the power interruption tests, which were performed at 15.5 kV, are shown in Table 1.1.

Table 1.1 Power interruption tests of SS5-1-1

RMS current interrupted (kA)	0-5	5-10	10-15	15-20	20-25	25-30	over 30
Number of tests	7	2	3	8	3	4	6

The switch was a completely sealed device but was designed to be opened up and refurbished. The effects of the aforementioned testing is clearly evident in Figure 1.8, which shows the inside of the two halves of the interrupter before and after the tests. Noting that the gaps between the petals were almost closed in some places by the splatter of molten metal, it was decided to make these wider in the

Figure 1.8   Effects of arc erosion, contacts and shields before and after testing
(Courtesy of General Electric Co.)

pair of contact to be used in the next series of tests. These tests were conducted in May 1959; Figure 1.9 shows a logbook page with the general layout of the test circuit and an oscillogram from a no load test, indicating contact travel and contact parting. SS5-2-1, as the refurbished 'Sputnik' was designated, surpassed itself in the test programme, the outcome of which is summarised in Table 1.2. The voltage for these tests was 15.5 kV, as before, but other tests were made at

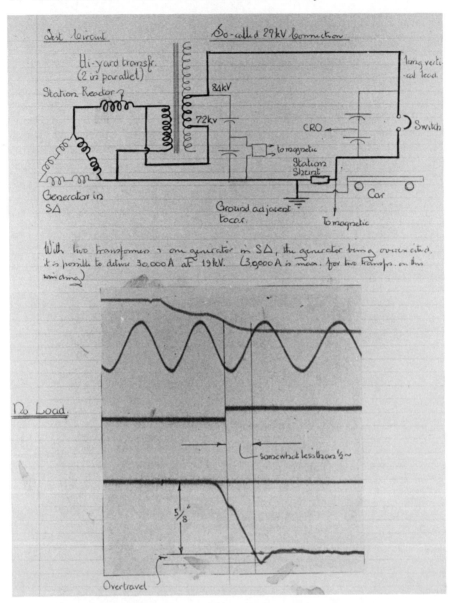

Figure 1.9   Page from author's log book, 20 May 1959, showing the test circuit and no load oscillogram for SS5-2-1

(Courtesy of General Electric Co.)

Table 1.2   Power interruption tests of SS5-2-1

RMS current interrupted	10-20	20-30	30-40	over 40
Number of tests	33	27	18	10

voltages ranging from 11.5 to 22 kV. Both symmetrical and asymmetrical currents were interrupted. In one particularly impressive 11.5 kV test, the RMS current was 54 kA; the switch interrupted a major current loop of 92 kA peak after 8 ms of arcing.

This test series revealed several other important facts:

(i) that a vacuum interrupter can support a very high rate of rise of recovery voltage following current interruption: 6 kV/μs was recorded

(ii) that dielectric integrity could be maintained following severe interruption tests - 110 kV DC was successfully applied across the open contacts

(iii) that vacuum interrupters can take some extraordinary abuse; after being pressed beyond its limit and sustaining two cycles of arcing before being relieved by the station back-up breaker, the device repeatedly performed to its prefailure level in subsequent tests.

'Sputnik' was an experimental interrupter, but it paved the way for commercial products which are discussed in Section 1.7. In the meantime I turn to concurrent events that were taking place in the UK.

1.6 The British connection

About the time that General Electric launched its vacuum programme, a more modest but extremely successful investigation was begun by the British Electrical and Allied Industries Research Association (ERA) in England. It was directed by M.P. Reece, to whom I am much indebted for supplying details of the effort. During the period 1953 to 1959, Dr Reece and his associates addressed a host of topics fundamental to the understanding of vacuum arcs and vacuum switching. The results of these many studies are detailed in three, originally confidential, ERA reports [19,20,21]. Subsequently, much of this material appeared in more condensed form in two pioneering papers [22,23].

The ERA work was both experimental and theoretical. The majority of the experiments were conducted with well-prepared electrodes (contacts) in the demountable system shown in Figure 1.10. This is very similar to the apparatus that my colleagues and I were using for similar purposes at about the same time.

Reece made much use of cinephotography to capture the structure of the vacuum arc. He measured the arcing voltage of different metal contacts and recognised how these were dependent on the thermal properties of the metals and drew far reaching conclusions on the choice of contact materials. The physical arrangement for a particularly insightful experiment [23] is shown in Figure 1.11. It was devised to measure the division of arcing voltage between the anode and the cathode by a calorimetric method. Two copper electrodes of equal mass and having plane parallel contacted faces were mounted in the vacuum chamber on two pieces of thin-walled nickel–silver tube. The temperatures of the two electrodes could be measured by means of thermocouples. After assembly in the vacuum chamber, preliminary arcs were run until the arcing voltage had settled to a steady value, and were continued for some time after to ensure that both electrodes were clean. The electrodes were then left in contact for some hours to cool and for their temperatures to equalise. An arc was then drawn with an arc length of about 2 mm, and after a convenient

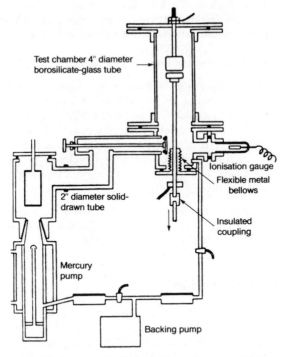

Test chamber 4″ diameter
borosilicate-glass tube

Ionisation gauge

Flexible metal
bellows

2″ diameter solid-
drawn tube

Insulated
coupling

Mercury
pump

Backing pump

Figure 1.10   Demountable vacuum system used for many of the ERA studies [19]

(Courtesy of Dr M.P. Reece)

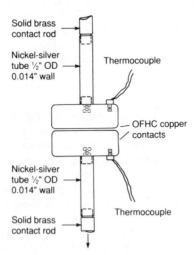

Solid brass
contact rod

Nickel-silver
tube ½″ OD
0.014″ wall

Thermocouple

OFHC copper
contacts

Nickel-silver
tube ½″ OD
0.014″ wall

Solid brass
contact rod

Thermocouple

*Figure 1.11   Contact arrangement for calorimetric measurement of division of arc voltage
between cathode and anode [23]*

(Courtesy of Dr. M.P. Reece)

charge had passed in the arc, the temperatures reached by the two electrodes were measured.

Three measurements were made at a current of 30 A, and in each case the ratio of the temperature rise of the anode to that of the cathode was 1.7:1. Thus, with a total arc voltage of 21 V, the effective voltage drop of the vacuum arc on copper is about 8 V, and the effective voltage drop is 13 V. After recording this result, Reece [22] made the following significant observation '. . . the effective anode voltage drop is absorbed, not near to the anode, but in the discharge very close to the cathode, where it does work which is stored in the kinetic energy and potential energy of moving particles, and which is transported to the anode by the movement of these particles and is liberated there.' As shown in Section 2.6, this observation is essentially correct. Reece used this data as a basis for energy balance computations at the electrodes.

In this theoretical analysis, Reece [23] computed particle velocities and hence the spatial and temporal densities of electrons, ions and neutral vapour. His calculations were very much of an engineering kind; by making simplifying but justifiable assumptions, he was able to obtain 'ball park' or reasonable estimates of important quantities. It was not science for science's sake, but engineering in the pursuit of vacuum switching technology. He identified what he believed to be critical parameters for current interruption and attempted to assess interrupting capability.

The work at the ERA did not encompass vacuum interrupter development (the budget was far too small for that) but a number of ideas were advanced for this development. One was the bimetal contact [24], which separated contact functions and thereby capitalised on the different characteristics of the two materials. These were later used in contactors. In addition, a great deal was learned which was subsequently used by British companies in later programmes. An outstanding example was the development of chrome/copper as a contact material [25]. This was invented by A.A. Robinson of the English Electric Company, but undoubtedly it was influenced by the material in Reference 19 on thermal properties of cathode materials, arc voltage and arc stability. Copper/ chromium remains today one of the best materials for vacuum interrupter contacts.

Reece joined Associated Electrical Industries (AEI) in 1961 and was the prime mover in that company's entry into the vacuum switchgear business. He was subsequently joined by Mitchell who undertook much of the basic experimentation. Of special note during this era was the introduction of the *contrate contact* [26,27] illustrated in Figure 1.12. This serves the same purpose as the spiral contact depicted in Figure 1.5, that is to say, it induces arc motion by magnetic forces when the arc is in the constricted mode and thereby disperses the heat flux to the contacts and reduces the production of metal vapour. Both the contrate contact and copper/chromium for the contact material were later adopted by Westinghouse when that company started its vacuum switchgear development.

Spurred on by AEI's success, a number of British competitors became active in the 1960s; these include English Electric, Reyrolle, and GEC. Towards the end of that decade the first two of these companies established a joint company for manufacturing vacuum interrupters which they named Vacuum Interrupters Ltd. (VIL). However, shortly after this, by acquisition and merger, AEI, English Electric and GEC became one company under the GEC banner. VIL continued

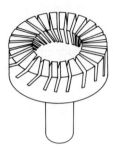

Figure 1.12   The contrate contact for producing arc rotation

to operate, producing a high quality product. In 1974 Reyrolle sold half its interest to the British company, thus, widening the base. The company operated this way until about 1983 when VIL became totally owned by GEC.

1.7 Penetrating a conservative market

The success of GE's vacuum recloser [15] in the US led to the introduction of more vacuum products, notably medium-voltage vacuum metal-clad switchgear in 1967. However, there was a well-entrenched alternative, the air magnetic circuit breaker. These were heavy and cumbersome devices that needed constant maintenance if operated frequently. Nevertheless, customers who were unfamiliar with the many benefits of vacuum – low maintenance, long life, flame-free quiet operation, etc. – were reluctant to try what they viewed as an unproved technology. This led to a Catch-22 situation in which the customer demanded a record of field experience before buying a new product.

About this time, General Electric made a decision to develop a full range of vacuum circuit breakers covering all voltage classes from 4 to 800 kV [28]. In 1967, transmission circuit breakers with ratings from 121–800 kV in general required the use of multiple interrupting modules in series to achieve the desired voltage performance. Mechanical, electrical and economic considerations pointed to a nominal three-phase voltage rating of 45 kV for such a module. Accordingly, the initial research and development efforts were directed to increasing the voltage capability beyond 15 kV.

By 1969, experimental interrupters had been tested to 20 kA at 45 kV, and a few years later a 40 kA, 45 kV module had been developed. Figure 1.13 illustrates that the basic construction followed the proven principles used for 15 kV interrupters. The contacts were of the spiral type (Figure 1.5) and opened to a gap of 3/4 in. The insulating metal enclosure was glass; it supported three metal-vapour condensing shields as opposed to the single shield for the 15 kV design. The overall dimensions were 8 in OD x 17 in long. The rating of this interrupter was

Three-phase system voltage	45 kV
Short-circuit current	40 kA
Asymmetry factor	1.3

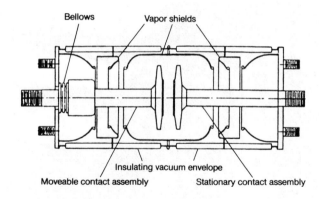

Figure 1.13   Outline drawing of a vacuum interrupter rated 45 kV, 40 kA [28]

(*Courtesy of General Electric Co.*)

Continuous current	2000 and 3000 A
Capacitance switching current	500 A
BIL	230 kV
Low-frequency withstand	100 kV

Circuit breakers incorporating these interrupters were designed, prototypes were built [29], but commercial products were never sold.

A similar situation evolved in the UK Like General Electric, AEI thought the future of vacuum lay in HV and EHV breakers. In 1967/68 they built and sold four 132 kV (145 kV max, design), 3500 MVA vacuum breakers, which corresponds to a short-circuit current rating of 16 kA. One of these is shown in Figure 1.14. They remained in service for many years, indeed, one is still believed to be operational at the time of writing (1991). AEI subsequently built a 5000 MVA prototype, but it never went into commercial service.

These high voltage breakers had too many parts; they did not capitalise on the best features of vacuum; they could not compete with the emerging SF_6 technology.

For the reasons just cited - resistance to change on the part of the utility industry and the misdirection of effort into a high voltage adventure on the part of the manufacturers - the position of vacuum switchgear in the first few years of the 1970s was less than encouraging. A new generation of equipment was needed if vacuum was to displace air magnetic breakers. The market requirements were well defined: more than 20 different ratings, improved reliability, smaller size, and a wide variety of customer options. Manufacturing's critical concern seemed to pull in precisely the opposite direction - more standardisation, fewer parts and drawings, moulded and punched parts - an echo of Henry Ford's 'any colour as long as it's black'.

These seemingly divergent requirements were reconciled by a feature of the vacuum interrupter that had received little attention during its development, and was of no significance to early investigators. This was the fact that a number of different voltage and current ratings could be accomplished by design changes *within* the interrupter, without grossly perturbing the *external* dimensions needed

Figure 1.14   British high-voltage vacuum circuit breaker rated 132 kV, 3500 MVA (circa 1967) located in West Ham, London

(Courtesy of Dr. M.P. Reece)

to interface with the rest of the circuit breaker. This allowed many different ratings to be offered in a family of breakers and equipment in which dimensions and designs could be standardised, and the variations in parts reduced with a consequential improvement in reliability, all without restricting the options offered to the market. The new breakers were approximately half the size of the older designs, permitting a two-tier arrangement in space where only one was possible previously.

The work at General Electric has been recounted for reasons of familiarity, but comparable programs were undertaken by other US manufacturers, most notably Westinghouse. The milestone contributions of Kimblin in the basics of vacuum arcs benefitted the entire industry. After a late start, Westinghouse, like General Electric, developed a very successful line of vacuum switchgear.

Parenthetically, it must be recorded that pioneers often have to endure crises which those that follow can avoid. There were several occasions in the early days of the General Electric programme when it was difficult for us who were involved to persuade a reluctant management to continue funding work which at times seemed slow in producing results. No such doubts assail the competition once it has been demonstrated that the job can be done. Money seems to come out of the woodwork in the game of catch-up.

The new breakers found wide acceptance in the late 1970s, especially by industrial (as distinct from utility) users, who found the space-saving feature particularly appealing. So it was that a few years later the major manufacturers in the US ceased to make air magnetic breakers; vacuum had become the industry's preferred choice.

1.8 The Japanese story

Japanese manufacturers played a substantial part in the design revolution just described. Entering the vacuum field later than the Americans and the British, they at first relied on imported technology. Vacuum switchgear is that kind of product; one can purchase the interrupters themselves and then adapt them to one's own design of breaker. Other electrical manufacturers (OEMs) discovered this in the US and built up thriving small businesses.

There seemed less reluctance on the part of the Japanese to experiment with new configurations. Moreover, their electric supply industry and their railroads, appeared more willing to accept vacuum switching devices. Designs proliferated as a consequence. These showed ingenious use of materials, often new materials, and often in dual functions such as insulation and mechanical support.

It was not long before Japanese manufacturers had set up facilities to make their own interrupters and once again a profusion of products emerged. Figure 1.15 gives an indication.

The Japanese were the first to commercially exploit the axial magnetic field as a means of improving power switching performance [30,31] (Section 4.4.3). Clever designs were produced to provide the axial field. An example by Kurasawa *et al.* [32] is shown in Figure 1.16.

Japanese engineers and scientists contributed much to our knowledge of vacuum arcs and the switching process and in the implications for power switching application. In this context, by way of example, I would mention the work of Murano and associates [33] on the inadvertent interruption of high-frequency currents during reignitions.

1.9 Concluding comments

At the outset of this chapter I said that in no way could this historical account be comprehensive. Perhaps some amends for omissions can be made by a few brief acknowledgments.

The major European manufacturers were late to join the vacuum club, but once convinced, they have made rapid progress in the technology, as is apparent from later chapters of this book. Their scientific and technical staff have

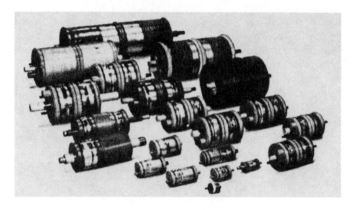

Figure 1.15   Japanese vacuum interrupters (circa 1978)

(Courtesy of Meidensha Electric Manufacturing Co. Ltd.)

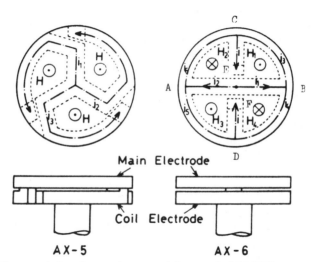

Figure 1.16   Contacts designed to produce an axial magnetic field [32]

advanced the art, sometimes with new techniques, such as the laser work of Schade, [34,35] (ABB).

Finally, one must acknowledge the thousands of contributions, worldwide, made by the personnel of universities and research institutes: Daalder in the Netherlands, Hantzsche, Juttner and Lindmayer in Germany, Fursey, Mesyats and Rakhovsky in the Soviet Union, Boxman and Goldsmith in Israel, Kutzner in Poland and Rondeel in Norway, to mention a few.

A special mention must be made of Professor Gunther Ecker of the University of Bochum, consultant, mentor and theoretician extraordinary to the global vacuum switching community. For almost four decades Dr Ecker has sponsored conferences, promoted discussion, and by his thoughtful and searching criticism has kept us honest.

1.10 References

1 AYRTON, H.: 'The Electric Arc' (D. Van Norstrand, New York, 1902)
2 COMPTON, K.T.: 'The Electric Arc', *Trans. AIEE*, 1927, **46**, pp. 868-883
3 SORENSEN, R.W., and MENDENHALL, H.E.: 'Vacuum switching experiments at the California Institute of Technology', *Trans. AIEE*, 1926, **45**, pp. 1102-1105
4 SORENSON, R.W.: 'The power application vacuum switch', *Electrical Engineering*, 1958, **77**, pp. 150-154
5 HOUSEKEEPER, W.G.: 'Sealing base metals through glass', *J. AIEE*, 1923, **42**, pp. 954-960
6 COBINE, J.D.: 'Research and development leading to the high-power vacuum interrupter — a historical review', *Trans. IEEE*, 1963, **PAS-64,** pp. 201-217
7 LINDVALL, F.C.: 'A general study of vacuum switch contacts of widely different materials when subject to abnormal duty', Doctoral thesis, California Institute of Technology, 1928
8 HAMILTON, J.H.: 'Measurement of arc voltage across open switch contacts', Doctoral thesis, California Institute of Technology, 1928
9 KLING, A.: 'A new vacuum switch', *General Electric Review*, 1935, **38**, pp. 525-526
10 KOLLER, R.: 'Fundamental properties of the vacuum switch', *Trans. AIEE*, 1946, **65**, pp. 597-604
11 JENNINGS, J.E., SCHWAGER, A.C., and ROSS, H.C.: 'Vacuum switches for power systems', *Trans. AIEE*, 1956, **75**, pp. 462-468
12 ROSS, H.C.: 'Vacuum switch properties for power switching applications', *Trans. AIEE*, 1958, **77**, pp. 104-117
13 SCHWAGER, H.H.: Discussion of Reference 12
14 LESTER, G.N., and PFLANZ, H.M.: Discussion of Reference 6
15 STREATER, A.L., MILLER, R.H., and SOFIANEK, J.C.: 'Heavy duty vacuum recloser', *Trans. AIEE*, 1962, **81**, pp. 356-363
16 SCHNEIDER, H.N.: 'Vacuum type circuit interrupter', US Patent 2 949 520, 1960
17 GREENWOOD, A.N., SCHNEIDER, H.N., and LEE, T.H.: 'Vacuum type circuit interrupter', US Patent 2 892 912, 1959
18 LEE, T.H., GREENWOOD, A.N., CROUCH, D.W., and TITUS, C.H.: 'Development of power vacuum interrupters', *Trans. IEEE*, 1962, **81**, pp. 629-639
19 REECE, M.P.: 'Vacuum switching I, review of literature and scope of ERA research on the subject', E.R.A. report G/XT166, 1959
20 REECE, M.P.: 'Vacuum switching II, extinction of an AC vacuum arc at current zero', ERA report G/XT167, 1959
21 REECE, M.P.: 'Vacuum switching III, experimental results, application to contactors, conclusions and bibliography', ERA report G/XT168, 1959
22 REECE, M.P.: 'The vacuum switch, part 1 — properties of the vacuum arc', *Proc. IEE*, 1963, **110**, pp. 793-802
23 REECE, M.P.: 'The vacuum switch, part 2 — extinction of an AC vacuum arc', *ibid*, pp. 803-811
24 REECE, M.P.: 'Improvements related to vacuum electric switches', British Patent 835 253, 1960
25 ROBINSON, A.A.: 'Vacuum type electric circuit interrupting devices', British Patent 1 194 674, 1970
26 REECE, M.P., and LAKE, A.A.: 'Improvements relating to vacuum switch contact assemblies', British Patent 1 098 862, 1968
27 KNAPTON, A.G., and REECE, M.P.: 'Improvements relating to vacuum switch contacts', British Patent 1 100 259, 1968
28 KURTZ, D.R., SOFIANEK, J.C., and CROUCH, D.W.: 'Vacuum interrupters for high voltage transmission circuit breakers', IEEE Conference Paper C75 054-2, 1975
29 SHORES, R.B., and PHILLIPS, V.E.: 'High voltage vacuum circuit breakers', *Trans. IEEE*, 1975, **PAS-94**, pp. 1821-1830
30 MORIMIYA, O., SOHMA, S., SUGAWARA, T., and MIZUTANI, H.: 'High current vacuum arcs stabilized by axial magnetic fields', *Trans. IEEE*, 1973, **PAS-92**, pp. 1723-1732
31 YANABU, S., SOHMA, S., TAMAGAWA, T., YAMASHITA, S., and

TSUTSUMI, T.: Vacuum arc under an axial magnetic field and its interrupting ability', *Proc. IEE*, 1979, **126**, pp. 313-320

32 KURASAWA, Y., KAWAKUBO, Y., SUGAWARA, H., and TAKASUMA, T.: 'Behavior of vacuum arcs in transverse magnetic field and axial magnetic field', Proceedings of 10th international symposium of Discharges and electrical insulation in vacuum, 1982

33 MURANO, M., FUZI, T., NISHIKAWA, H., NISHIKAWA, S., and OKAWA, M.: 'Voltage escalation in interrupting inductive current by vacuum switches', *Trans. IEEE*, 1974, **PAS-93**, pp. 264-280

34 GELLERT, B., SCHADE, E., and DULLNI, E.: 'Measurement of particles and vapor-density after high-current vacuum-arcs by laser techniques', Proceedings of 12th international symposium of Discharges and electrical insulation in vacuum, 1986

35 DULLNI, E., SCHADE, E., and GELLERT, B.: 'Dielectric recovery of vacuum after strong anode spot activity,' Proceedings of 12th international symposium of Discharges and electrical insulation in vacuum, 1986

Chapter 2

The vacuum arc

2.1 Definition

Since the vacuum arc is the key element in the vacuum interrupter, some knowledge of its structure and behaviour is essential to understanding the operation of vacuum switchgear. The name vacuum arc is really incorrect, indeed, it is a contradiction. As Ecker [1] points out 'If there is a vacuum there is no arc, and if there is an arc there is no vacuum'. A more exact name would be metal vapour arc, since the arc which forms when current-carrying contacts separate in a vacuum 'burns' in the metal vapour of the contacts. The name vacuum arc is now so pervasive and so uniformly accepted that I use it throughout this book as meaning a metal vapour arc.

Like arcs in gaseous media, a vacuum arc, once established, is relatively stable, or as we say, it is self sustaining. It will continue to operate, drawing its energy from the electrical system, until some external agency, such as the occurrence of a current zero, removes its source of energy. This statement is not completely accurate in that at small currents the arc can become unstable and spontaneously extinguish. This phenomenon known as *current chopping*, is discussed in Section 4.5.

One may therefore define a vacuum arc as a self-sustained electrical discharge, between two electrodes, which maintains current flow by vapour and charge carriers derived from one or both of the electrodes. The next four sections are largely descriptive, that is to say, they describe the appearance of the arc, discuss the arc voltage and note how the arc responds to magnetic fields. Subsequently, at the end of the chapter, I attempt to explain these observations, at least qualitatively, by means of a model for the vacuum arc.

2.2 Appearance of a vacuum arc

Vacuum arcs have more than one 'mode' and their appearance depends on their mode. The mode in turn depends on the current level and to a significant extent on the size of the contacts. At lower currents the arc assumes the *diffuse mode*. This is characterised by one or more exceedingly bright spots on the cathode (negative) electrode. These *cathode spots* are in constant motion over the contact surface like skaters on a rink, and appear to repel each other, even to the extent of sometimes moving down the side of the contact. They have a finite, though variable lifetime; new ones are created, often by the splitting of existing spots, as other spots extinguish. The number of spots is determined by the magnitude of the current; each spot on copper carries of the order of 100 A. Thus, a copper vacuum arc of 1,000 A will have approximately ten cathode spots. In an AC halfcycle, the number of spots increases progressively to the peak of the current and then decreases until, near current zero, only one is left. The current per spot varies from one cathode material to another [2,3] being highest for refractory materials and lowest for low boiling point metals. The remainder of the diffuse arc discharge contrasts sharply with the cathode spots in that it has a much paler luminosity. The name diffuse is apt, for most of the interrupter is filled with glowing luminous plumes of conical shape with the apex of each pointing towards its cathode spot; the anode contact is bathed in this glow. The entire assemblage looks like, and behaves like, a number of independent, parallel arcs. It reminds one of a watering can, where the plasma issues from small 'holes' on the cathode surface like the jets of water from the sprinkler. As the current is increased, the luminous plasma expands until it fills almost the entire volume of

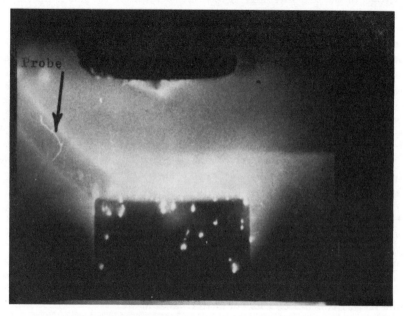

Figure 2.1   Vacuum arc in the diffuse mode [4]

(Courtesy of Dr G.R. Mitchell)

the vessel. Figure 2.1 gives an overall impression of a diffuse vacuum arc. This figure was provided by Mitchell, who has given a comprehensive description of the vacuum arc [4,5].

An examination of the surface of the cathode after it has supported a diffuse arc reveals faint tracks, popularly known as hen tracks, over its surface. A closer examination with higher magnification shows these tracks to be a criss-cross of craters and other debris. Harris [6] graphically describes such remains as '. . . like the track of a herd of cattle along a muddy trail'. It is evident that the surface in such areas has been turbulently molten. The surface of the anode electrode or contact has a matt appearance following diffuse arcing. Metal, vapourised at the cathode spots, travels radially away to strike and condense on neighbouring objects. Since the anode usually subtends a large angle for most of the cathode, the majority of this material condenses on the anode. Some vapour escapes from between the contacts and freezes on the shield. In switches which have glass envelopes and glass shields, this material is clearly evident as a ring of deposited metal, encompassing the shield directly across from the contact gap.

As the current in the vacuum arc is increased still further, there is a sudden and remarkable change in its appearance. The plasma, instead of bathing the anode in the manner just described, becomes focused on a small area of that electrode (perhaps 1 cm^2). This *anode spot*, which is usually at a sharp edge of the contact, is clearly molten; it is playing an active part in the discharge. The small cathode spots become grouped together, giving the arc a much brighter columnar appearance which is referred to as the *constricted mode*. This is evident in Figure 2.2 which, like Figure 2.1 was obtained by high speed movie photography.

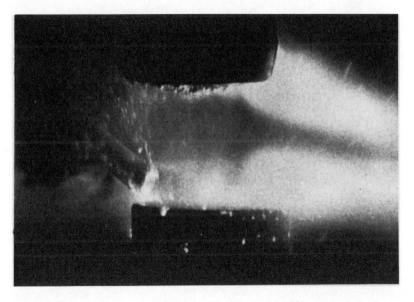

Figure 2.2   Constricted vacuum arc between 1 in diameter OFHC copper electrodes at 2 cm separation at 15 kA showing anode (upper) and cathode (lower) columns [4]

(Courtesy of Dr G.R. Mitchell)

The current at which the transition from diffuse to constricted mode takes place depends on the contact material. It is also very dependent on the size of the contact. This has been very well demonstrated by Kimblin [7], who also showed that the transition current decreases as the length of the arc is increased and/or the size of the anode is reduced. Rich [8], on the other hand, has built switches with very large contact areas and has thereby maintained the arc in the diffuse mode for peak currents up to 72 kA.

The constricted mode of the vacuum arc produces considerable erosion of both

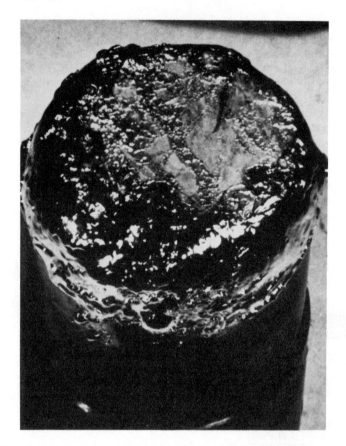

Figure 2.3   Severe erosion of a 1 3/4 in diameter OFHC copper anode after 10 loops of 50
Hz current, with peak of 17.5 kA and at 1 cm contact separation [4]

(Courtesy of Dr. G.R. Mitchell)

contacts, the amount depending on the magnitude and duration of the current, or the charge in coulombs conveyed by the arc, but being one or two orders greater than the cathode erosion of the diffuse arc. We are no longer simply talking of evaporation from the cathode: gross melting occurs, especially at the anode. As current increases, jets of vapour emerge from the molten areas. This is clearly evident in Figure 2.3. Very high pressure gradients create veritable gales which drive droplets from the molten surface and deposit them as splatter on the shields. Such debris is clearly evident in Figure 1.8 which shows the interior of an interrupter after protracted high current arcing. I have on occasions peeled off lacy appendages, perhaps 4 cm^2 in area, from the surface of a contact. These were the residual spray fragments from the concluding milliseconds of a high current discharge, frozen as the arc was spent.

Later in this chapter, and more especially in Chapter 4, current interruption is seen to be much easier to achieve in the absence of anode spots than it is when they are present. The transition from diffuse to constricted mode of the arc is therefore of considerable practical significance. The onset of the transition is signalled by changes in the arc voltage which is therefore the subject of the following section.

2.3 Arc voltage

To anyone acquainted with the voltage across gaseous arcs, the voltage of a vacuum arc is surprisingly low. For a 200 A copper arc, for example, the arc voltage is about 20 V. Even lower arc voltages have been measured for other metals, calcium having 13 V, and somewhat higher values for refractories, 26 V for molybdenum, according to Davies and Miller [9]. Observe that I have made no mention of the length of the arc. This is because the length has little influence when the arc is in the diffuse mode. The voltage is concentrated across the cathode region; the electric field gradient in the diffuse plasma cones which extends across most of the gap, is very small.

There is a certain amount of high-frequency noise on the arc voltage, about 2 V or thereabouts is typical. Its frequency spectrum ranges from kilohertz to megahertz. The magnitude of this noise increases significantly as the current level is reduced, at 10 A, for instance, excursions of 60 V can be observed [10]. Closer examination reveals that it comprises unidirectional pulses (short-lived positive spikes). It is my experience that arcs of all kinds have an incredible will to survive (combating this idiosyncracy has provided job security for many switchgear engineers over the years). This line of thought suggests that these pulses are momentary adjustments to maintain emission in the cathode spot.

As the current is increased the voltage increases slowly. Again, this is in direct contrast to the gaseous arc which has a negative resistance characteristic, voltage falling as current increases. It is the positive characteristic of the vacuum arc which allows many cathode spots to operate simultaneously in the same discharge (and for individual vacuum interrupters to be operated in parallel).

Still further increase in current brings a considerably greater increase in arc voltage, the average value going up by 40 V or more in copper arcs. Average is used because, again, there is much noise present, excursions may exceed 100 V. This activity is the precursor to the transition from the diffuse to the constricted mode of the discharge described in the last section. Once an anode spot is established, the arc voltage often drops and always becomes much less noisy. Figure 2.4 illustrates the different conditions just described as the arc current increases and passes through the transition from diffuse to constricted mode. It reveals something else, what might be described as a hysteresis effect, whereby the arc voltage, having dropped with the formation of an anode spot, does not increase again later in the halfcycle when the current is once more reduced. Mitchell [5] shows some excellent evidence of this.

At this stage the vacuum arc has more of the characteristics of a high pressure arc, which indeed it has become. The channelling of current into a relatively small molten area of the cathode creates considerable erosion, as noted. A jet of metal vapour, at or close to atmospheric pressure is the consequence.

Miller [1] describes other modes of the vacuum arc which essentially occur during the transition from the diffuse to the constricted mode. From a vacuum

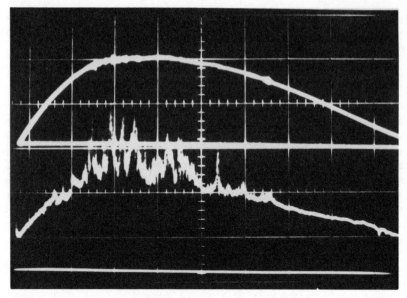

Figure 2.4   Arc current (upper trace) and arc voltage (lower trace) for 7.5 cm diameter OFHC copper electrodes at 0.5 cm contact separation [4]
Scale: 1 div. = 10.5 kA = 25 V = 1 ms

(Courtesy of Dr G.R. Mitchell)

switchgear point of view, these do not have any practical significance and for this reason they are not discussed here.

2.4   The vacuum arc in a magnetic field

In discussing how vacuum arcs are affected by magnetic fields, one must differentiate between transverse and axial magnetic fields. In the transverse case, the magnetic field is perpendicular to the principal direction of current flow in the arc. A radial magnetic field, as produced by two opposing coils, coaxial with the axis of a switch, would satisfy the first condition. If the current in one of these coils is reversed so as to produce a reinforcing magnetic field, the arrangement would satisfy the second condition for an axial field. Consider the transverse or radial field first and confine interest to observation without any attempt to interpret what is observed.

When the arc is in the diffuse mode it is seen to move when a transverse magnetic field is applied. It moves in a direction mutually perpendicular to the field and the current. However, and this is very important, instead of the $\mathbf{J} \times \mathbf{B}$ relationship with which we are familiar for the force on a current-carrying conductor in a magnetic field, what is observed is a $-\mathbf{J} \times \mathbf{B}$ effect, i.e. the cathode spot moves in a direction *opposite to* the normal amperian motion. Such translation is referred to as *retrograde motion*. Clearly, the diffuse column of the vacuum arc is a conductor which we would expect to move as would any other conductor when a transverse magnetic field is applied. One can only conclude that there is a superior opposing force experienced by the spot itself which accounts for the anomaly.

Gallagher [12] has studied retrograde motion and discovered that it is affected by the background pressure. If this is raised by introducing a gas into the vacuum, the speed of the spot motion is reduced. At a critical pressure (of the order of 1 torr for mercury) the direction of motion reverses.

When there are several cathode spots operating simultaneously (as in Figure 2.1), the magnetic field of one parallel arc will be experienced by all others. The retrograde forces will be such as to cause one spot to repel and be repelled by all the others, a fact that is readily observed in practice.

If the arc current is increased beyond the point where the arc changes from diffuse to constricted mode, the motion of the arc in a transverse magnetic field is no longer retrograde but reverts to the classical $\mathbf{J} \times \mathbf{B}$. The cathode spots are observed to bunch and the current becomes concentrated at the anode. Both these changes are consistent with magnetic pinch forces, i.e. with $\mathbf{J} \times \mathbf{B}$ action.

The application of an axial magnetic field (AMF) has several noteworthy effects [13] which are very important from a practical point of view. First, there is a distinct change in the appearance of the arc. For an arc of several thousand

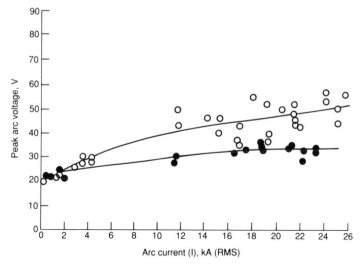

Figure 2.5   Peak arc voltage against RMS arc current for interrupter with electrode spacing at voltage peak of approximately 0.4 cm [13]

   o *no applied field*
   • $B = 2 \times 10^{-2} \ T/kA$

amperes, there will be, say, 20 cathode spots. In the absence of the field, the plasma plumes from these spots will diverge and mix to engulf the anode. High speed photography shows that the application of an AMF causes each discharge to become more columnar and distinct, so that the assemblage appears like so many separate and parallel arcs.

A second important consequence of the AMF is a sharp reduction in the arc voltage. Kimblin's data [13], illustrating this effect, is shown in Figure 2.5. Section 2.2 noted that there is an increase in arc voltage as the arc current is increased and that this is a presage of anode spot formation. One might therefore suppose that the application of an AMF, with the concomitant reduction in arc voltage, would delay anode spot formation. This is indeed the case, the arc in a given vacuum interrupter can be maintained in the diffuse mode up to a much higher current when an AMF is present than it can when it is not. Diffuse arcs are easier to interrupt than constricted arcs with anode spots, thus, the AMF has practical implications for vacuum switchgear (Chapter 4).

2.5   Positive ions in the vacuum arc

The very low arc voltage of the vacuum arc strongly suggests that most of the contact gap wherein the arc is burning is filled with a neutral plasma. This requires the presence of positive ions in about the same concentration* as the electrons. The presence of such ions has been confirmed experimentally by Plyutto *et al.* [14] and by Davies and Miller [9]. What is surprising, however, is the energy of the ions. It has been repeatedly found that the average ion energies correspond to voltages exceeding the total arc voltage. This discovery caused much consternation among researchers. It is a fact to consider in the next section where the subject of vacuum arc modelling is addressed.

2.6   Model for the vacuum arc

2.6.1   Introduction

A model is a physical and/or mathematical description of some thing or some process, that helps us to better understand the object or what is going on in the process. A good model explains what we observe physically. It should be consistent with all our observations, or as a minimum, it should not be in conflict with any of them. In the best of circumstances it should be predictive, that is to say, it should accurately foretell what to expect if this or that parameter is changed.

A model for a vacuum arc could be useful in that it is impossible to measure directly with any degree of accuracy, many important features of the arc, the dimensions, current density and temperature of the cathode spot, being good examples. One can only deduce these from more indirect observations. A model can draw from many other sources - experiments, verified theories, etc. - knowledge which can be put into, and tested out in, a model, so that in favorable circumstances, there emerges a *modus operandi* which accords with all or most that we know about the object or process.

What are some of the observations that a model of a vacuum arc should explain? These are manifold, but some of the more important ones are:

- why the current concentrates at small spots on the cathode
- why the mode of the arc changes from diffuse to constricted
- why the arc is influenced as it is by magnetic fields

---

*If the ions are multiply charged, there would, of course, be fewer.

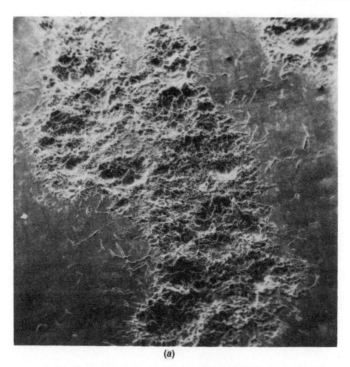

(a)

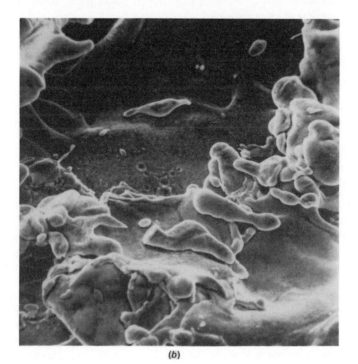

(b)

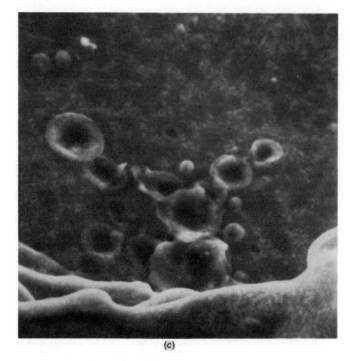

(c)

Figure 2.6   Scanning electron micrographs of cathode tracks from vacuum arc on copper cathode at magnifications:

(a) *X50*, (b) *X1000*, (c) *X5000* [16]

(*Courtesy of Dr L. P. Harris*)

- why the arc voltage has the values observed and why it varies as it does
- why the arc is a source of relatively high-energy ions.

It is helpful to begin with a low current arc and focus attention initially on the cathode.

2.6.2   *Model for the cathode spot*

The cathode spot is the minute area in which current issues from the contact surface and the similarly minute volume of intense glow adjacent to the current emission site. Spots of this kind are clearly identifiable in both Figures 2.1 and 2.2. They appear much bigger than they are because of the way the light diffuses in the photographic emulsion and also because, even during the short time of the exposure, they move.

From the work of many people, particularly Daalder [15], it appears that the spot on the surface is about 20 m in principal dimension. This figure has been arrived at by examining the fossil remains left on the cathode surface after the arc has extinguished. Harris [16] shows some excellent photographs of such craters, taken by scanning electron microscopic: Figure 2.6 is an example. These are the sources which are in continual motion and are continually splitting; new ones are created and old ones die throughout the duration of the arc.

There is no universal agreement that the spots are homogenous. Kesaev [17] believes, and some others are inclined to agree with him, that each little area comprises a number of subcomponents or cells, perhaps ten or so, which are the basic elements. Given the dimension just cited and the observed value of approximately 100 A per cathode spot, it is evident that the current density is of the order of 10^{11}–10^{12}A/m^2. Such a high current density combined with the resistivity of the contact, causes considerable joule heating within the contact, behind the spot: this alone will raise the local temperature. The cathode surface in the spot area is also being bombarded by positive ions, which adds greatly to the local energy input. As a consequence, the cathode is boiling, a stream of metal vapour is issuing from the surface. There is a quite intense electric field in this region caused by positive ion space charge. The combination of high temperature and strong field create conditions for a considerable flux of electrons to be emitted from the electrode surface at the cathode spot.

The subject of electron emission from metals is a profound one. It has been the cause of exhaustive inquiry over many years and the source of hundreds, if not thousands, of papers. Helpful discussions of the subject will be found in References 18 and 19. The topic is of sufficient importance to the understanding of vacuum arcs to make a diversion here for a brief summary.

According to the band theory of metals, electrons in great numbers (about 8.5×10^{22}/cm^3 for copper) in the conduction band of the metal move freely through the crystal lattice. They endow the metal with its high electrical conductivity inasmuch as a considerable current can be caused to flow by the application of a very small electric field. However, for electrons to escape from the metal, even the most energetic of the population at the top of the conduction band, i.e. at the Fermi level, they must overcome a barrier of several volts (~3.8 eV for copper), known as the *work function*. Essentially, no electrons have this kind of energy at room temperature, so electrons do not escape from metals unaided (except to other conductors) at room temperature.

When a metal is heated, on the other hand, the average energy of the conduction band electrons increases in proportion to the temperature rise. The distribution of energies is Gaussian, about the most likely energy. Thus, with increasing temperature, some fraction will acquire sufficient energy to overcome the work function and escape to the surrounding space. This process, which is referred to as *thermionic emission*, was first reported at length by Richardson [20] and generally quantified by Dushman, who gave the following relationship between current density, J(A/m^2) and temperature, T(°K):

$$J = AT^2 \exp\left[\frac{-\phi_0 e}{kT}\right] \tag{2.6.1}$$

where ϕ_0 is the thermionic work function, k is Boltzmann's constant (1.37×10^{-23} joules/K) and e is the electronic charge. The constant A is approximately 6×10^5 for most metals. It is apparent that for $\phi_0 = 3.8$ eV and $T = 3,000$K

$$J \simeq 2 \times 10^6 \text{A/m}^2$$

which is exceedingly small on the scale discussed for cathode spots. It is evident from this rough calculation that only refractory metals, such as tungsten and

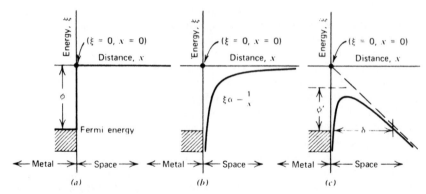

Figure 2.7   Potential energy against distance for electron near metal surface [16]

(a) *Without image charge field*
(b) *With image charge field*
(c) *With image charge and external field*

(*Courtesy of John Wiley & Sons*)

molybdenum can produce high thermionic emissions. Metals like copper and silver would melt and boil before reaching a high enough temperature.

Electrons can also be pulled out of metals by a strong electric field, a process appropriately known as *field emission*. The field has the effect of reducing the potential barrier. This can be understood from Figure 2.7 (after Farrall [18]). Figure 2.7(a) shows conditions at the ideal boundary between the metal (on the left) and the vacuum (on the right). Inside the metal, the electrons are confined within the conduction band (the cross-hatched region) which extends down from the Fermi level. The Fermi level itself is ϕ volts below the surface potential, where ϕ is the work function. It is usual to consider the metal surface as being at zero potential, thus, the Fermi level is at $-\phi$ volts. To escape from the metal, Fermi electrons must gain ϕ eV of energy.

In point of fact, an electron outside of the metal but in close proximity to it, disturbs the field because of its charge. The form of the potential profile, which can be determined from considering the field of the electron and its image within the metal, is shown in Figure 2.7(b). A uniform electric field in the space to the right would produce the linear change of potential indicated by the dotted line in Figure 2.7(c). The combined potential profile can be obtained by superposition, as indicated by the full line in Figure 2.7(c). Note how the barrier has been reduced by the external field and how it will be reduced more if the field is increased. Electrons can escape from the metal by passing over the barrier, which becomes easier as the barrier is reduced.

According to quantum mechanics, there is a finite probability that an electron within the metal will *tunnel through* the potential barrier. This probability is markedly dependent on the width of the barrier, which, for Fermi energy electrons, is shown as δ in Figure 2.7(c). The probability increases as the width decreases, or put another way, the probability increases with increasing field.

The field emission process just described pertains to cold electrodes subjected to a strong electric field. It was first established in quantitative terms in 1928 by

Fowler and Nordheim [21]. Their expression for the current density is

$$J = \frac{0.01541 E^2}{\phi t^2(y)} \exp\left[\frac{-6.831 \times 10^9 \phi^{3/2} v(y)}{E}\right] \text{ A/m}^2$$

$$y = \frac{3.795 \times 10^{-3} E^{1/2}}{\phi}$$

(2.6.2)

where E is the electric field (V/m) and ϕ is the work function. The functions $t(y)$ and $v(y)$ have been tabulated by Miller [22]. They are slowly varying functions of E and ϕ, and as such, are frequently treated as constants.

It is a relatively straightforward matter to determine whether or not a set of experimental data (I against V) fits the Fowler–Nordheim equation. The procedure is to rewrite Equation 2.6.2 as

$$\log_{10}\left[\frac{J}{E}\right] = -\log_{10}\left[\frac{\phi t^2(v)}{1.541 \times 10^{-2}}\right] - \frac{6.831 \times 10^9 v(y) \phi^{3/2}}{2.3026} \frac{1}{E}$$

(2.6.3)

To a reasonable approximation, bearing in mind what has been said about the functions, this is of the form

$$\log_{10}\left[\frac{J}{E}\right] = A + \frac{B}{E}$$

(2.6.4)

where A and B are constants. Thus, if $\log_{10}[J/E]$ (or $\log_{10}[J/E]$, assuming a constant emitting area) is plotted against $1/E$, the data should follow a straight line. Many plots of this kind have revealed the expected linear relationship, confirming the validity of the theory.

Experiments made with very sharply pointed tungsten emitters using a field emission microscope [23], permitted J and E to be determined quite accurately, since the geometry of the emitting point was known accurately. These tests showed the magnitude of J to be as predicted by the Fowler–Nordheim equation over the range of E tested. This situation did not prevail, however, for tests made on large area electrodes, where a number of emission sites were probably active simultaneously. The observed emission was much greater than theory predicted. This could be accommodated by introducing a field enhancement factor, designated β. If the macroscopic field was determined to be E, the microscopic field due to surface asperities etc. was βE. Unfortunately, less than credible values were obtained for β (often 100 or more), which put the geometric interpretation in doubt. This problem remains unresolved. There is a general agreement that it is very difficult to determine the size of emitting areas (and hence J from I), and that emission sites are much smaller than had been previously supposed.

It was stated that the cathode surface at the location of a cathode spot is both very hot and under the influence of a strong electric field. One would expect, therefore, that one would enhance the other as far as electron emission is concerned. Shottky [24] was the first to consider the combined effect of temperature and field; his paper appeared in 1923. It was not until many years later that Murphy and Good [25] put the subject into a rigorous framework and described quantitatively what is now known as *T–F* emission (*T* for temperature, *F* for field).

With the foregoing background on electron emission from metals, discussion of a model for the cathode spot can be resumed.

Lee and Greenwood [26] in 1961 were the first to invoke *T–F* emission as the active emission mechanism in the cathode spot of a vacuum arc. In a paper described by Ecker [27] as 'The first attempt at a comprehensive treatment of the vacuum arc spot . . .' and '. . . the first step forward in the understanding of the vacuum arc', they proceeded to set down basic equations relating to the dependent variables in the cathode region. One, for example, was concerned with energy balance at the cathode surface, another was the space charge equation. With the aid of these equations and certain critical limiting conditions (there could be no more ions produced than neutrals evaporated) and the

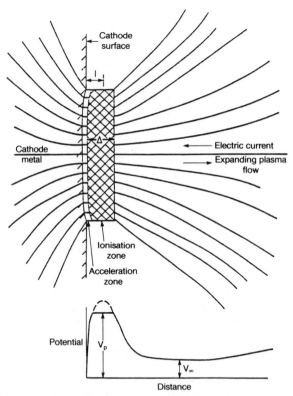

Figure 2.8   Cathode cell geometry and potential distribution in Harris model [28]

physical properties, thermal conductivity, evapouration constants, work function, etc., of the cathode, they were able to predict arc behaviour, particularly with respect to current chopping. This matter is discussed in more detail in Section 4.5. The state of knowledge has increased greatly since that time and so has the rigor of theoretical analysis. For these reasons, Harris' model [28] is used to improve understanding of cathode spots. A brief summary of this work is presented, but hopefully sufficient to understand the critical processes involved and their dependence on the material properties of the cathode.

The cathode spot model is shown pictorially in Figure 2.8. On the left is the cathode itself, on the right, the plasma. In the middle is the ionisation zone, shown crosshatched. Axial symmetry is assumed, so that the cathode spot is circular, as is the ionisation region. It is vitally important to understand the scale of this diagram. As indicated earlier, the cathode spot diameter on copper is ~20 μm. The distance ℓ to the centre of the ionisation zone is the mean free path (MFP) for ionisation of the copper vapour. Because the local pressure is very high (many atmospheres) the MFP is very short, of the order of $1-10\times10^{-8}$m [29].

Thus, the ionisation zone, as Mitchell and Harris describe it [27], '. . . has the proportions of a very thin pancake'.

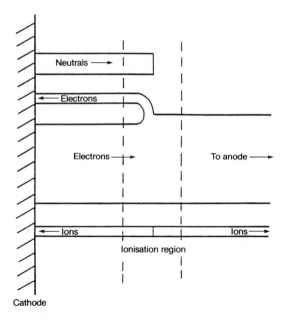

Figure 2.9   Particle fluxes in cathode region of vacuum arc [27]

Vapour boiled off from the cathode, and electrons emitted from the same incandescent surface, interact energetically in the ionisation zone. Essentially all the neutrals are ionised, many of them doubly and some of them triply [9]. The ionisation zone is a place of particle generation and therefore of high local pressure since pressure is directly proportioned to particle concentration ($p = nkT$). Neutral atoms, electrons and ions enter the zone from the cathode, and plasma flows out in both directions. The fluxes of the different species are shown in Figure 2.9, where the width of flows provides a measure of their relative strengths [27]. This diagram has been constructed on the assumption that all the neutrals suffer single ionisation, thus, the neutral stream has the same strength as the sum of the ion streams. Because of the pronounced differences in mass and mobility between electrons and ions, the ionisation region becomes positively charged and the electric potential in this region exhibits a local maximum, a 'potential hump,' as indicated in Figure 2.8.

As Mitchell and Harris [31] point out, the hemispherical expanding plasma flow from the ionisation region toward the anode provides an essentially neutral conducting medium that spans most of the interelectrode gap and permits the passage of electric current with only small voltage drop. The plasma flow from the ionisation zone toward the cathode provides both an intense energy flux and a high space charge field at the cathode surface, and consequently strong emission of both neutral atoms and electrons. The emitted atoms and electrons flow away from the surface, across the acceleration zone, to the ionisation zone, where they mix by collisions to feed both energy and particles into the plasma.

From Figure 2.9 note that in the acceleration zone a fraction of the current designated s is carried by electrons. The ion current is therefore $(1-s)I$. In the expanding plasma there is also a component $(1-s)I$ of ion current, but note that this is flowing in opposition to the main current, thus the electron current in this region is $(2-s)I$. Neutrality can be preserved by adjustment of ion and electron anode-directed velocities. The current density can be written

$$J = nev \qquad (2.6.5)$$

where n and v are the concentration and velocity of the electrons or ions as the case may be. To preserve neutrality

$$n_+ = n_-$$

or

$$\frac{J_+}{V_+} = \frac{J_-}{V_-}$$

assuming the ions are singly charged. Whence,

$$\frac{V_-}{V_+} = \frac{J_-}{J_+} = \frac{(2-s)}{1-s} \qquad (2.6.6)$$

The electrons and ions completely mingle, but the effect is like two trains, one of electrons and one of ions, moving in the same direction but at quite different speeds on the same track. Since this is somewhat difficult to visualise, think instead of the two trains as travelling on parallel tracks. Recourse to this analogy will be made again when current interruption is discussed in Chapter 4.

The anode-directed ion flow is the source of the high energy ions observed by Davis and Miller [9] at points quite remote from the cathode surface. The ions acquire their energy by ambipolar expansion of the ionisation zone plasma. Because of the great disparity of the masses of the electrons and ions, the energy of the flow, which in the ionisation zone resides mainly in ionisation energy and thermal motion of electrons, is converted to kinetic energy of directed flow of the ions by ambipolar diffusion.

This model description is largely qualitative, but Harris [28] provides the analytical rigor to support his model. He treats the cathode itself behind the cathode spot, the surface of the cathode spot, the acceleration zone, the ionisation zone, the expanding plasma and the linkages between them. For example, at the cathode surface, the energy flows due to ion bombardment, electron and atom emission and thermal conduction are balanced. Emission of electrons from the surface is calculated and the adequacy of this calculation is monitored by a consistency check. Evaporation of atoms from the surface is calculated from the vapor pressure characteristics.

In the acceleration zone, charge and particle numbers are conserved, forces and energies are balanced, and development of electric fields by excess positive space charge is accounted for in an approximation of Poisson's equation.

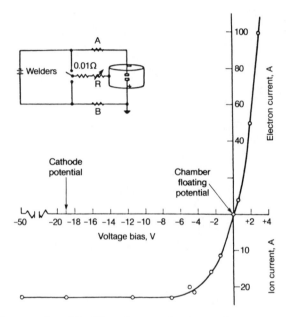

Figure 2.10 *Current collected by 27cm-diameter chamber wall as function of voltage bias from 'floating potential'* [32]

> *Arc current 275 A; electrode spacing 1.3 cm; 2.5 cm-diameter electrodes*
> *Insert: experimental circuit*
>
> (*By permission of IEEE*)

In the ionisation zone, charge and mass are conserved, and forces and energy fluxes are balanced. Near electrical neutrality is maintained in this region. The distinctive features of this zone are the net creation of charged particles by ionisation, and compressibility effects in the resulting high temperature plasma.

2.6.3 Model for the arc column and the anode

The picture drawn so far of diffuse vacuum arc has the interelectrode space more or less filled with a luminous plasma with the anode acting as a passive collector of electrons to satisfy the needs of the external circuit. Section 2.3 pointed out that for short arcs there is almost no voltage across the plasma, virtually all of the arc voltage is concentrated across the cathode fall. Kimblin [32] has performed some enlightening experiments on the diffuse arc. When the metal vapour shield is floating (its normal operating mode) it behaves like a Langmuir probe [33] and for copper electrodes, takes up a potential of approximately +19 V with respect to the cathode. Kimblin [32] measured the current flowing to the shield as its potential was varied over a considerable range. The results are shown in

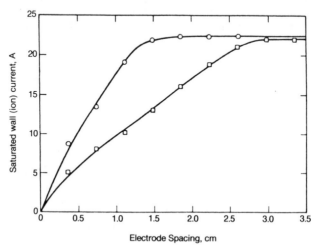

Figure 2.11   Wall ion current as a function of electrode spacing for electrode diameters of 2.5 and 5 cm [32]

Arc current 275 A; chamber biased to cathode potential

○ = 2.5 cm-diameter electrodes
□ = 5 cm-diameter electrodes

(By permission of IEEE)

Figure 2.10, where shield potential is measured relative to what would be its floating value. Observe that even a modest positive potential draws a considerable electron current from the plasma. Likewise, a negative bias attracts ions, but the ion current saturates. In some of the experiments being reported, the 'shield' was in fact the cylindrical wall of the vacuum vessel which the anode and cathode penetrated by bushings.

Kimblin [32] used this arrangement to investigate ion current as a function of arc current, electrode spacing, electrode diameter, and wall diameter. The ion current dependence on electrode spacing and electrode diameter for a 275 A arc is shown in Figure 2.11, where the data points represent the mean ion current observed for several arcing sequences as the given electrode spacing. For 2.5 cm diam. electrodes, the ion current first increases linearly with electrode spacing and reaches a maximum saturated value for a spacing of ∼1.5 cm. For 5 cm diam. electrodes, the ion current approaches the same maximum but at a longer electrode spacing of ∼3 cm. A similar dependence of the ion current on electrode separation and diameter is observed throughout the current range 100 to 3000 A, although the ion currents attain maximum values at slightly smaller electrode spacings with increasing arc current.

The maximum ion current values vary from 8% of the total arc current at 100 A to 20% at 3000 A, as shown by the data points of Figure 2.12. The □ symbol represents the maximum ion current to the 5.7 cm diam. shield of the interrupters, and the ○ symbol represents the maximum ion current to the

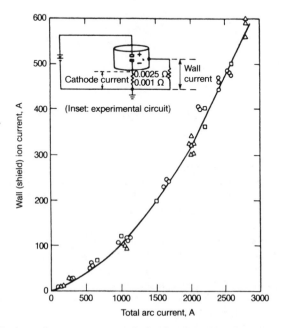

Figure 2.12   Maximum ion current to negatively biased wall against total arc current (sum of 'cathode current' and 'wall current') [32]

    ○ *Wall diameter 27 cm and* □ *shield diameter 5.7 cm (electrode spacing 1.3 cm; 2.5 cm diameter electrodes)*

    △ *Wall diameter 27 cm (electrode spacing 3 cm; 5 cm diameter electrodes)*

    (By permission of IEEE)

27 cm diam. wall for identical 2.5 cm diam. electrodes. The maximum ion current is obviously independent for the collecting wall diameter, and thus one can infer that practically zero plasma ionisation takes place away from the immediate interelectrode gap. The points designated △ in Figure 2.12, show the maximum ion current drawn to the wall for 5 cm-diam. electrodes. From the good agreement with the smaller electrode data, it is apparent that the maximum ion current from a vacuum arc of a given current is a fundamental property, independent of the collecting wall diameter and the electrode diameter.

Any model devised for the interelectrode plasma must be consistent with the results of the experiments just presented. Section 2.6.2 described the plasma as

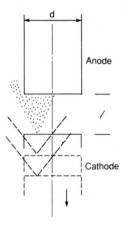

Figure 2.13   Separating contacts with plasma plume from cathode spot

an expanding plume emanating from the cathode spot. Reece [34] writes of a conical plume with a halfangle of about 30°. The cathode spot or spots are moving about the cathode surface and sometimes pass down the side, thus, a crude representation of the plasma might be that shown in Figure 2.13. This shows cathode and anode with plasma between, some of which is escaping the gap and reaching the shield.

The dotted profiles show sequential positions that the cathode might occupy as it is retracted downward during the opening of a switch. Now review Figures 2.10 to 2.12, in light of Figure 2.13. As ℓ increases, more plasma escapes to be collected at the shield. Thus, the magnitude of the ion current increases with increasing gap as shown in Figure 2.11. What fraction of the plasma passes to the shield and what is collected by the anode depends on the solid angle subtended by the anode. The smaller the electrodes for a given gap, the smaller will be this solid angle and the greater will be the shield ion saturation current, which accords with the two curves of Figure 2.11. But why does the ion current saturate? In the Harris model for the cathode [25], described in Section 2.6.2, a fraction of current $(1-s)$ is carried by positive ions in the plasma. Once this fraction has been drawn to the shield by negative bias, there is no more to attract. Kimblin's observed figure of $s = 0.08$ for a single cathode spot accords with Harris' analysis [28] and is also close to the fraction which Lee and Greenwood [26] found necessary for energy balance at the cathode. As the plasma expands away from the cathode spot, the ion and electron concentrations diminish approximately as the inverse square of the distance from the cathode spot. This tends to reduce the conductivity. Mitchell [35] refers to the effect of vapour 'starvation'. When a pair of current-carrying contacts are separated a point may be reached in their travel when the arc voltage must increase for the anode to attract sufficient flux of electrons to satisfy the circuit requirements. This will occur at a shorter gap for smaller diameter electrodes, since, as noted already, it is the ratio ℓ/d that determines the angle subtended by the anode at the cathode. The voltage gradient will increase close to the anode where electrons will tend to predominate to produce an anode fall.

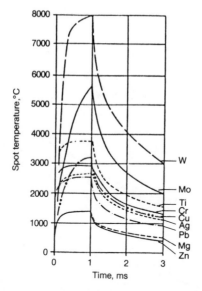

Figure 2.14   Thermal response of materials to arc heating $3 \times 10^9 W/m^2$ [28]

(*By permission of IEEE*)

The formation of electron space charge near the anode and the concomitant creation of an anode fall has implication for the anode itself. The energy of the electrons striking the anode will be measured by the energy they acquire as they are accelerated through the anode fall. They already have considerable thermal energy acquired in the ionisation region and they also deliver ϕ eV by way of heat of condensation, ϕ being the anode work function. With high currents and/ or small anode areas, the heat flux to the anode is such that its temperature rises quite rapidly. Rich [8] and Mitchell and Harris [28] have computed the temperature rise for a specific heat flux to a semi-infinite plane; the latter's results are reported.

Initially when the heat pulse is applied, the surface temperature will rise at a rate which varies inversely with the thermal conductivity of the electrode material, until surface evapouration losses become a comparable cooling mechanism. If the heat input is sustained and/or increased, the surface temperature quickly reaches a 'plateau' which is close to the boiling point of the electrode material. Figure 2.14 shows the response of an anode surface to a heat input of $3 \times 10^9 W/m^2$, for a period of 1 ms. There will be occasion to consider the consequences of different materials at other points in this text, but for the moment we concentrate on copper as a frequently used material for contacts in vacuum switchgear. The presence of metal vapour and the flux of electrons passing through it are potent agents for producing more plasma by impact ionisation. If this occurs, the positive ions will tend to neutralise the electron space charge resulting in a condition that can lead to an instability. The appearance of a new source for electrons and ions causes local elimination of the cathode fall. The current therefore concentrates and focuses in the region where ionisation of anode vapour develops, and the arc voltage drops as a consequence.

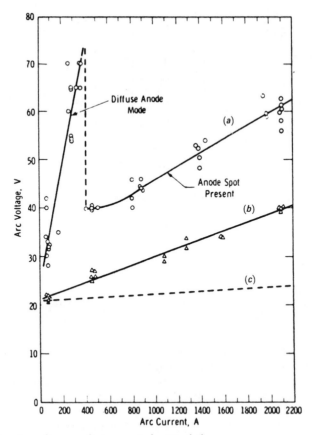

Figure 2.15   *Arc voltage against current characteristics*

(*a*) *2.5 cm-electrode spacing (1.3 cm-diameter anode)*
(*b*) *2.5 cm-electrode spacing (5 cm-diameter anode)*
(*c*) *Minimum electrode spacing (both anodes)*

(*By permission of IEEE*)

What we are describing is the formation of an anode spot, mentioned in Section 2.2, with the accompanying reduction in arc voltage described in Section 2.3. Such spots are prone to develop near the edge of a contact where the ability of heat to soak into the anode is reduced by the geometry. The appearance of an anode spot signals the transition from the diffuse to the constricted mode of the discharge. Magnetic pinch forces help maintain the current constriction at the anode and volumes of vapour and molten droplets pour into the gap in a jet from the anode surface.

This chapter concludes with graphical evidence to illustrate the mode transition; again, it is due to Kimblin [7,32]. Figure 2.15 shows the arc voltage against arc current for two sizes of electrode. Curve (a) is for 1.3 cm-diameter electrodes, curve (b) for 5 cm-diameter electrodes. Note how the spot forms at a quite low current on the smaller anode. Curve (c) represents the arc voltage when the contacts first separate.

The sketch of the vacuum arc provided in this chapter is far from complete, but hopefully it serves to help understand how vacuum switching devices function, and explain their various idiosyncrasies.

2.7 References

1 ECKER, G. *in* LAFFERTY, J. M. (Ed.): 'Vacuum arcs – theory and application' (Wiley, 1980) p. 229
2 DJAKOV, B. E., and HOLMES, R.: 'Cathode spot division in vacuum arcs with solid metal cathode', *J. Phys. D*, 1971, **4**, pp. 504–509
3 KIMBLIN, C. W.: 'Erosion and ionization in the cathode spot regions of vacuum arcs', *J. Appl. Phys., 1973*, **44**, pp. 3074–3081
4 MITCHELL, G. R.: 'The high current vacuum arc and its relevance to circuit interruption', PhD thesis, London Polytechnic, 1967
5 MITCHELL, G. R.: 'High-current vacuum arcs, part 1 – an experimental study', *Proc. IEE*, 1970, **117**, pp. 2315–2326
6 HARRIS, L. P. *in* LAFFERTY, J. M. (Ed.): 'Vacuum arcs – theory and application' (Wiley, 1980) p. 136
7 KIMBLIN, C. W.: 'Anode voltage drop and anode spot formation in DC vacuum arcs', *J. Appl. Phys.*, 1969, **40**, pp. 1744–1752
8 RICH, J. A.: 'A means for raising the current for anode spot formation in metal vapor arcs', *Proc. IEEE, 1971*, **59**, pp. 539–545
9 DAVIS, W. D., and MILLER, H. C.: 'Analysis of the electrode products emitted by DC arcs in a vacuum ambient', *J. Appl. Phys.*, 1969, **40**, pp. 2212–2221
10 Reference 6, p. 154
11 MILLER, H. C.: 'Discharge modes at the anode of a vacuum arc', Proceedings of 10th international symposium on Discharge and electric insulation in vacuum 1982
12 GALLAGHER, C. G.: 'The retrograde motion of the arc cathode spot', *J. Appl. Phys.*, 1950, **21**, p. 768
13 KIMBLIN, C. W., and VOSHALL, R. E.: 'Interruption ability of vacuum interrupters subjected to axial magnetic fields,' *Proc. IEE*, 1972, **119**, pp. 1754–1758
14 PLYUTTO, A. A., RYZHKOV, V. N., and KAPIN, A. T.: 'High speed plasma stream in vacuum arcs', *Socv. Phys. JETP*, 1965, **20**, p. 328
15 DAALDER, J. E.: 'Diameter and current density of single and multiple cathode discharges in vacuum', *IEEE Trans.*, 1974, **PAS–93**, pp. 1747–1757
16 Reference 6, pp. 137–145
17 KESAEV, T. G.: 'Stability of metallic arcs in vacuum', *Sov. Phys., Tech. Phys.*, 1963, **8**, p. 447
18 FARRALL, G. A. *in* LAFFERTY, J. M. (Ed.): 'Vacuum arcs – theory and application' (Wiley, 1980) pp. 21–24
19 COBINE, J. D.: 'Gaseous conductors' (McGraw–Hill, 1941) pp. 106–122
20 RICHARDSON, O. W.: 'Emission of electricity from hot bodies', *Proc. Cambridge Philos. Soc.*, 1901, **11**, p. 286
21 FOWLER, R. H., and NORDHEIM, L.: 'Electron emission in intense electric field', *Proc. Royal Soc.*, 1928, **119**, p. 173
22 MILLER, H. C.: 'Values of Fowler-Nordheim field emission functions, $v(y)$, $t(y)$, and $s(y)$', *J. Franklin Inst.*, 1966, **282**, p. 382
23 MÜLLER, E. W.: 'Electronenmicopische Beobactungen von Feldkothoder', *Z. Phys.*, 1937, **106**, p. 541
24 SHOTTKY, W.: 'Emission of electron from an incandescent filament under the action of a retarding potential', *Ann. Phys.*, 1914, **44**, p. 1011
25 MURPHY, E. L., and GOOD, R. H.: 'Thermionic emission, field emission and transition region', *Phys. Rev.*, 1956 **102**, p. 1464
26 LEE, T. H., and GREENWOOD, A.: 'Theory for the cathode mechanism in metal vapor arcs', *J. Appl. Phys.*, 1961, **32**, pp. 916–923
27 Reference 1, p. 255
28 HARRIS, L. P.: 'A mathematical model of cathode spot operation', Proceedings of

8th international symposium on Discharges and electrical instalation in vacuum, Albuquerque, NM, USA 1978
29   Reference 19, p. 20
30   GREENWOOD, A., CHILDS, S.E., MODY, H.K., SULLIVAN, J.S., FARRALL, G.A., VAN NOY, J.H., FROST, L.S., GORMAN, J.G., and KIMBLIN, C.W.: 'Fundamentals of interruption in vacuum', US Department of Energy report DOE/ET/29197-1 (DE86007974), 1984
31   MITCHELL, G. R., and HARRIS, L. P.: 'The structure of vacuum arcs and the influence on the design of vacuum interrupters', IEEE winter power meeting, conference paper C75 067-4, 1975
32   KIMBLIN, C. W.: 'Vacuum arc ion currents and electrode phenomena', *Proc. IEEE*, 1971, **59**, pp. 546–555
33   Reference 17, p. 135
34   REECE, M. P.: 'The vacuum switch part 1 – properties of the vacuum arc', *Proc. IEE*, 1963, **110**, pp. 793–802
35   MITCHELL, G. R.: 'High-current vacuum arcs part 2 – theoretical outline', *Proc. IEE*, 1970, **117**, pp. 2327–2332

Chapter 3

Vacuum breakdown

3.1 Implications for vacuum switchgear

The beginning of the last chapter makes the point that some knowledge of the vacuum arc is essential if one is to understand vacuum switchgear; the same is true for breakdown in vacuum. The literature on this subject is enormous, but much of it is not directly relevant to present concerns. I refer, for instance, to the work on the carefully prepared, finely polished electrodes of fixed gaps. The electrodes in a vacuum interrupter may be the contacts, which are frequently brought briskly together, experience welding, are pulled apart and suffer arcing. Alternatively, the electrodes could be shields, end plates and other hardware forming part of the vacuum enclosure. These parts may have irregular geometries, they may be joined by dielectric surfaces and they may be sprayed by metal vapour and/or particles when the switch is operated. These influences cause frequent changes in surface conditions which can dramatically change their ability to support high potential differences between the affected parts. In short, conditions within a vacuum interrupter are not pristine and moreover, are constantly changing.

A particularly critical time for breakdown is the interval that follows arc interruption when, hopefully, the contact gap is recovering its dielectric strength to hold off the transient recovery voltage that the circuit is about to impress across it. A breakdown at this time, or a reignition as we would call it, would signal a failure to interrupt. Immediately after current zero it cannot be said that a vacuum exists, since some residual plasma from the prior arcing is still present in the contact gap. Moreover, parts of the contacts may still be hot, perhaps even molten, and certainly capable of producing more vapour. The interrupter structure may also be suffering mechanical vibration at this time, as a consequence of the impact the mechanism provides for breaking contact welds or, if the contacts are near the end of their stroke, the action of buffers designed to arrest their motion. During this period, which typically lasts for a few microseconds, we surely face a very dynamic situation as far as breakdown is concerned.

These considerations mean that from a breakdown point of view, the behaviour of a vacuum interrupter is likely to be more statistical and less deterministic than, say, a gas blast circuit. Accepting this fact, it is still possible to learn much from theory and from fundamental experiments with vacuum gaps. An excellent general survey of this subject, by Farrall, will be found in Chapter 2 of Reference 1.

3.2   Basic notions about breakdown in gas and breakdown in vacuum

Some knowledge of the kinetic theory of gases is necessary is to understand breakdown in gases. The unfamiliar reader would be well advised to refer to a good text on the subject, of which there are many [2,3]. A brief summary of the kinetic theory is sufficient for the present purpose.

Under normal circumstances, according to the kinetic theory, the atoms or molecules of a gas are in random motion and are colliding frequently with each other and with the walls of the vessel. They have no preferred direction, all directions are equally probable. The mean velocity of the particles is dependent on their kinetic energy which in turn depends on the temperature of the gas. The kinetic energy is divided equally between the degrees of freedom, each having $(1/2)kT$, where k is Boltzmann's constant $(=1.37\times10^{-23}$ joule/K) and T is the absolute temperature (K). The total kinetic energy per cubic metre of a monatomic gas is therefore $(3/2)nkT$, where n is the particle concentration per m^3. For a gas at NTP (0°C and 760 mm Hg) $n=2.71\times10^{25}$, thus,

$$\text{total kinetic energy} = 3 \times 2.71 \cdot 10^{25} \times 1.37 \times 10^{-23} \times 273$$
$$= 3.041 \times 10^5 \text{ joule}$$

A diatomic gas has five degrees of freedom, the three translational degrees just described, plus one vibrational between the atoms and one rotational for the molecule as a whole. Its total kinetic energy is therefore $(5/2)kT$. For a monatomic gas

$$\frac{mC^2}{2} = \frac{3}{2}kT \tag{3.2.1}$$

where m is the mass of the atom (kg) and C is the RMS velocity (m/s), or

$$C = \left(\frac{3kT}{m}\right)^{1/2} = 6.411 \times 10^{-12}\left(\frac{T}{m}\right)^{1/2} \tag{3.2.2}$$

For argon $(m=6.637\times10^{-27})$ at NTP, $C=1,300$ m/s. The velocities of the individual particles in a gas follow the Maxwell–Boltzmann distribution

$$\frac{dN_c}{N} = \frac{4}{\sqrt{\pi}}\left(\frac{m}{2kT}\right)^{3/2} C^2 \exp-\frac{mc^2}{2kT}$$

which gives the fraction of the total population N having velocities between C and $C+dC$. This is an asymmetrical bell-shaped curve about the most probable velocity, $C_0 = (2kT/m)^{1/2}$ as indicated in Figure 3.1. At the fringes of the distribution some atoms have much higher velocities than C_0 and some have much less.

As mentioned, the particles make frequent collisions with each other and with the walls of the vessel. These collisions are elastic, that is to say energy and momentum are conserved during the encounters.

There are always present in the gas particle population a small number of

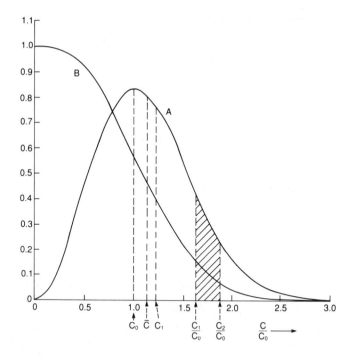

Figure 3.1   Maxwellian velocity distribution functions

Ordinate of A: $\dfrac{C_0 dN_c}{N dc}$

Ordinate of B: $\displaystyle\int_{\frac{c}{c_0}}^{\infty} \dfrac{dN_c}{N}$

charged particles (electrons and ions). They are the product of ionisation brought about by radiation which is ever present from background radio-activity, cosmic rays, etc. These are shortlived, they recombine, usually in the presence of a third particle or at the wall. Like the neutrals, the charged particles are in random motion, indeed, charged and neutral particles are in equilibrium.

If an electric field is applied, this equilibrium is disturbed. The charged particles experience a directed force which accelerates them in the direction of the field, thereby adding to their kinetic energy. The energy from the field, acquired by electrons and ions between collisions, depends directly on the field strength and the distance they travel between collisions, i.e.

$$\Delta KE = eEd = \frac{m}{2}\left(v_2^2 - v_1^2\right) \tag{3.2.4}$$

where e is the electronic charge, E is the field, and d the distance travelled. The velocity gained, $(v_2 - v_1)$ is much less for ions than for electrons because of the great difference in their masses. To put in some numbers, if $E = 500$ kV/m, and

$d = 10^{-6}$m, the energy transferred to the particle is

$$5 \times 10^5 \times 1.6 \times 10^{-19} \times 10^{-6} = 8 \times 10^{-20} \text{ joules}$$

this would bring a stationary electron to a velocity

$$\left[\frac{2\Delta KE}{m_e} \right]^{1/2} = 4.19 \times 10^5 \text{ m/s}$$

or a positive argon ion,

$$\left[\frac{2\Delta KE}{m_i} \right]^{1/2} = 4.91 \times 10^3$$

It is more usual to measure particle energy in *electron volts* (eV), one electron volt being the energy acquired by an electron in rising through a potential difference of 1 volt. Thus, 1 eV $= 1.602 \times 10^{-19}$ joule.

The distance travelled between collision varies widely about an average which is referred to as the *mean free path* (MFP). As one would expect, the MFP varies inversely as the particle concentration, n (particles per m^3) and the particle size. The distribution of free paths about the mean is given by

$$n = N_e^{-x/L} \tag{3.2.5}$$

or the number of free paths of length greater than a given distance is a decreasing exponential function of the distance. This indicates that only 37% of the initial number of particles have free paths greater than the MFP.

As the strength of the electric field is increased, the charge carriers acquire more and more energy, to the point where those particles with long free paths will gain sufficient energy to ionise neutrals with which they collide. That is to say, the collision will be inelastic, an electron will be removed from the impacted atom, leaving a positive ion behind. The energy required varies with the gas atom involved, Table 3.1.

Table 3.1   Ionisation potentials

Gas	N	O	H	A
Ionisation potential	14.48	13.55	13.5	15.69 ev

Electrons produced in this way will themselves be accelerated and can also engage in ionisation collisions. When such a process occurs it is aptly described as an avalanche and is associated with Townsend [4] who first investigated the phenomenon. Figure 3.2 gives an impression of such an avalanche. Townsend [3] introduced a coefficient α to define the number of electrons produced in the path of a single electron travelling a distance of 1 cm in the direction of the field.

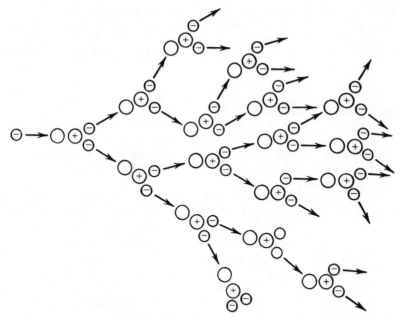

Figure 3.2   Electron avalanche in a gas

The current i then flowing in the gaps is

$$i = I_0 \epsilon^{\alpha d} \qquad (3.2.6)$$

where d is the gap length (cm) and I_0 the current at the cathode electrode. In nonuniform fields, as the voltage is increased, the process is first evident in the strong field region. In a point-to-plane gap, for example, it will occur close to the point. At atmospheric pressure we see a bluish glow around the sharp edges of electrodes, or hardware in general, when high potential differences exist. Such discharges are referred to as *corona*. What we are observing is the process of *excitation* rather than ionisation. This occurs with a lower energy than ionisation, 6.3 eV for the first excitation level of nitrogen. An electron is raised to a higher orbit in the atom by the energy received from the impacting particle. The visible radiation is the photons (light) emitted when the electrons of excited atoms return to their ground state.

In relatively uniform fields, an avalanche, when it develops, may spread completely across the gap as a *spark discharge*; we say that breakdown has occurred; the voltage across the gap collapses. More generally, conditions for breakdown require that the process described by Equation 3.1.6 become regenerative; the process must replenish itself by seeing to it that fresh initiating electrons are produced near the cathode. Under these conditions the expression for current becomes [5]

$$i = I_0 \frac{\epsilon^{\alpha d}}{1 - \gamma(\epsilon^{\alpha d} - 1)} \qquad (3.2.7)$$

where the coefficient γ accounts for the secondary processes which replenish the initiating electrons. This could be by bombardment of the cathode by positive ions produced by the avalanche, or by photons from the discharge causing electron emission. Note that Equation 3.2.7 reduces to Equation 3.2.6 when $\gamma = 0$. In the event that

$$\gamma(\epsilon^{\alpha d} - 1) = 1 \tag{3.2.8}$$

the denominator of eqn. 3.2.7 becomes zero and the current becomes infinite. This is deemed to be the condition for breakdown.

Precisely what happens in any circumstances depends also on the external circuit. If this has a high impedance, the current is limited to a low value; if the impedance is low, a power arc develops.

Changing the gas pressure affects the events just described. An increase in pressure increases the gas density, i.e. the particle concentration, which reduces excitation and ionisation. The result is a rise in the breakdown voltage. Conversely, reducing the pressure reduces the breakdown voltage. For a uniform gap between clean polished electrodes, and at atmospheric pressure, the breakdown field is approximately 30 kV/cm. Inside an airblast breaker, the gas pressure surrounding the open contacts is typically 600 psi (or 41 atmospheres). This is sufficient to hold off several hundred kV though the separation between live parts is only a few centimetres.

At the other end of the scale, when the pressure in argon is reduced to the order of 10 torr, a field of 50 V/m will cause breakdown and a steady glow discharge with a current of 30 mA can be maintained by a voltage of 500 V across electrodes 10 cm apart. The current is controlled by an external impedance. With adaptation, such discharges are used for illumination.

What is the situation if we continue to reduce the pressure? Clearly, the MFP of the gas will continue to increase, so the energy acquired by charged particles will increase. On the other hand, the number of such particles is reduced as the entire population is reduced, and the number of collisions is reduced for the same reason. We would therefore expect that for a given gap geometry the breakdown voltage would decline to some minimum value as the pressure is reduced and then increase again with further pressure reduction. This is precisely what Paschen [6] observed in his classic experiments more than 100 years ago and from which he enunciated in his famous law. Figure 3.3 shows an example of a Paschen curve with its minimum, which happens to be for hydrogen. The material of the electrode also influences the curve. What is of importance is the product pd, that is the pressure and the gap length or, more accurately, the density and the gap length, since pressure varies with temperature. We recognise that density controls the number of collisions in a unit volume and that gap length controls the volume. The breakdown voltage is unchanged if a reduction in density is compensated by a corresponding increase in gap length, which is essentially what Paschen discovered.

In the normal course of events the ambient pressure inside a vacuum interrupter assures that the device operates well up the left-hand (low pd) section of the Paschen curve. The characteristic length of gaps between the parts of an interrupter is of the order of 1 cm. Thus, if the pressure is 10^{-7} torr, $pd = 10^{-6}$ torr mm, which is beyond the extreme limit of Figure 3.3, what, therefore, is the

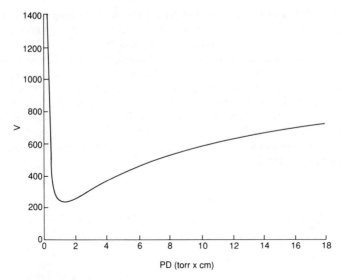

Figure 3.3   Paschen curve for hydrogen

relevance of the material of the last several paragraphs to the operation of a vacuum interrupter? It was pointed out in Section 3.1 that for a very short period of time following interruption of a vacuum arc, plasma from the prior arcing remains in the gap, as does metal vapour from one or both contacts. The plasma disperses very quickly, but the decline of metal vapour density is slower since hot spots on the contact surfaces take time to cool and therefore continue to emit vapour.

As a consequence the pressure may be much higher than the vacuum prevailing under quiescent conditions. The interrupter is vulnerable to breakdown during this brief time window. Moreover, a sparkover is more likely to occur across a 10 cm gap (from end to end of a switch) rather than a 1 cm gap (between the contacts), everything else being equal. This is because on this part of the Paschen curve the breakdown voltage falls as *pd* increases. Everything else may not be equal, there may be a higher density of vapour between the contacts. Nevertheless, long path breakdown must always be considered a possibility.

3.3   Electrode effects

If we separate the contacts of a vacuum interrupter to establish a gap of a few millimetres and then apply an increasing voltage (AC or DC) to the gap, we find that breakdown occurs and that the first breakdown may be at a quite low stress, perhaps 10 kV/mm or less. Since in this situation, the mean free path is much greater than any dimension of the vacuum vessel, the breakdown and conduction mechanisms are no longer possible. This means that charge carriers must be generated in some other way; an obvious possibility is the electrodes. Section 2.6.2 discusses the various ways in which electrons can be emitted from metals. These are thermionic emission from very high temperature cathodes (a process restricted to refractory materials), field emission where a powerful electric field is present at the cathode surface, and *T–F* emission, a combination of temperature and field, as occurs in the cathode spot of vacuum arcs on copper, silver or similar metals. It is not clear how these sources of emission apply to the experiment just described where breakdown is initiated by a modest voltage impressed across the open gap of a vacuum interrupter. There is no high temperature and apparently no powerful field. Moreover, if electrons are emitted, there is nothing in the gap for them to ionise. Some clues begin to emerge if the experiment is continued, i.e. if one reapplies voltage after breakdown has occurred.

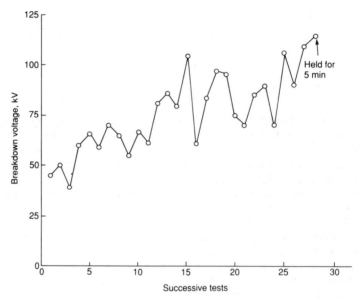

Figure 3.4   Successive breakdowns of contact gap of a vacuum interrupted, showing contact conditioning, obtained by the author in 1960

I have conducted tests of this kind many times and what one finds is that breakdowns continue to occur, but at steadily increasing voltage. Some typical data are displayed in Figure 3.4. There is a certain amount of scatter in the points and that they appear to be approaching some asymptotic limit. This is because the surge impedance of the test circuit was only 8Ω, thus a breakdown at 100 kV could deliver 12.5 kA (at 62.5 kHz). The scatter is less if such an experiment is conducted by placing high resistance, 500 k–5 MΩ, in series with the high-voltage source, so that breakdowns when they occur conduct only a small current, much of it coming from the discharge of local stray capacitance. The procedure just described is referred to as conditioning, or sometimes spark cleaning. If a much lower resistance is used, so that breakdown creates an arc, a certain amount of 'deconditioning' occurs, i.e. when voltage is reapplied, breakdown occurs at a relatively low value again. Deconditioning will also occur if the contacts are mechanically closed and then opened again.

These observations suggest that contact surface conditions affect breakdown in these circumstances. The belief is that on typical contacts, even new ones, there are many asperities and occlusions, the last-mentioned may be metallic or not conducting, and that these are the source of greatly enhanced electron emission. Scanning electron microscopic images confirm the presence of such excrescences [7]. Grain boundaries also appear to be important emission sites. Fowler–Nordheim plots, on the lines described in Section 2.6.2, strongly suggest that field emission is the source of electrons under these conditions, however, the results imply that field enhancement factor β is very high, or the work function at emitting sites is unusually low. It would also seem from these calculations that the emitting sites themselves are very small.

There are two theories with respect to the mechanism of breakdown. The first proposes joule heating of a sharp asperity on the cathode, which causes it to vapourise and then be ionised by the electrons. The second theory invokes anode involvement. It proposes that an electron beam, emanating from a site on the cathode, focuses on a spot on the anode and imparts to it a heat flux to cause local anode vapourisation. It is likely that both of these mechanisms are operative, the first prevailing in short gaps and the second in longer gaps. Both theories accord with the phenomenon of conditioning, since both have the effect of removing asperities from the contacts. With repeated breakdowns, a higher field is required to obtain breakdown conditions with the remaining surface irregularities. Closing the contacts, or creating an arc between them, creates a new set of debris from weld residue or crater remains.

Miller and Farrall [8] describe an ingenious experiment that clearly demonstrates the dominance of the cathode in the conditioning process. They utilised three electrodes, but used them in pairs, any one could be an anode or cathode. Having conditioned one pair, the third was used to replace the first anode; the result was inconsequential. When the experiment was repeated and the third electrode was used to replace the conditioned cathode, breakdown voltage was greatly reduced and a fresh sequence of conditioning had to be undertaken to restore the high breakdown voltage integrity.

Much evidence over a wide range of voltage and gap length suggests that breakdown voltage in vacuum varies as the square root of the gap length (see, for example Trump and van de Graff [9] and Anderson [10]). Cranberg [11] postulated his clump theory to account for this relationship. 'Clump' refers to a loosely adhering particle on the surface of one or other of the electrodes. The presence of a field in the gap implies surface charge of opposing polarities on the electrodes, some of which charge is shared by the clump. If the clump should break loose, it would accelerate across the gap and deliver an impact to the opposite electrode. The amount of energy delivered will depend on the mass of the clump and the square root of the gap voltage. It was Cranberg's proposal that when this energy exceeds some critical value, breakdown would result. The mechanism for the breakdown, whether it is vapourisation of a part of the impacting bodies, or as Olendskaya [12] and Chatterton and Biradar [13] have proposed, a consequence of a discharge between the clump and the opposing electrode just before impact, was not disclosed. There is no question that free particles, metallic and nonmetallic, exist within the envelopes of vacuum interrupters and that these are a cause of breakdown. On a number of occasions I have seen evidence of this. For example, during the conditioning of a demountable interrupter, I found that striking the stand on which the equipment was mounted frequently precipitated a breakdown.

Much of what is discussed in this section leads one to expect that contact material will have a significant effect on vacuum breakdown since different metals have quite different work functions, vapour pressure curves, hardness, brittleness, Young's modulus, electrical and thermal conductivities, etc., all of which appear to be influencing factors in one way or another. Experiments by Rozanova and Granovskii [14], Bouchard [15] and McCoy et al. [16], have confirmed this, yet the information is not necessarily very helpful from a practical point of view as far as vacuum switchgear is concerned. This is because breakdown is but one of several important considerations in selecting contact materials. Other requirements may be in direct conflict.

There are, of course, other stressed gaps in a vacuum interrupter besides the one between the contacts. Specifically, note shield-to-shield and shield-to-end plate gaps. Here there may be more latitude in selecting material. Nickel and stainless steel, for example, have proved good materials for shields.

Besides sparkover of gaps, one must also consider flashover of insulating surfaces. It is well known from the work of Kofoid [17], that placing an insulator in parallel with a vacuum gap greatly reduces the insulating strength of the gap. In the first place, there is a concentration of electrical stress at the triple point where the electrode, insulator and vacuum meet. Also, the field is often distorted by surfaces charges on the insulation. Gleichauf [18] found that the breakdown voltage was independent of degree of vacuum over a considerable range, but observed that the breakdown voltage increased with the surface resistivity of the insulator.

3.4 References

1 LAFFERTY, J. M. (Ed.): 'Vacuum arcs – theory and application' (Wiley, 1980)
2 LOEB, L. B.: 'The kinetic theory of gases' (McGraw–Hill, 1927)
3 COBINE, J. D.: 'Gaseous conductors' (McGraw–Hill, 1941)
4 TOWNSEND, J. S.: 'Electricity in gases' (Clarendon Press, Oxford, 1914)
5 MEEK J.M., and CRAGGS, J.D.: 'Electrical breakdown of gases' (Clarendon Press, Oxford, 1953)
6 PASCHEN, F.: *Wied. Ann.*, 1889, **37**, p. 69
7 LITTLE R. P., and WHITNEY, W.T.: 'Electron emission preceding electrical breakdown in vacuum', *J. App. Phys.*, 1963, **34**, pp. 2430–2434
8 MILLER H. C., and FARRALL, G.A.: 'Polarity effect in vacuum breakdown electrode conditioning', *J. App. Phys.*, 1965, **36**, pp. 1338–1344
9 TRUMP J. G., and VAN DE GRAFF, R. J.: 'The insulation of high voltage in vacuum', *J. App. Phys.*, 1947, **18**, pp. 327–332
10 ANDERSON, H. W.: 'Effect of total voltage on breakdown in vacuum', *Electrical Engineering*, 1935, **54**, pp. 1315-1320
11 CRANBURG, L.: 'The initiation of electrical breakdown in vacuum', *J. App. Phys.*, 1952, **23**, p. 518
12 OLENDZKAYA, N. F.: 'Vacuum breakdown with transfer of conducting particles between electrodes', *Radio Eng. & Electron. Phys.*, 1963, **8**, p. 423
13 CHATTERTON P.A., and BIRADAR, P.I.: 'Microparticle processes occurring prior to vacuum breakdown', *Z. Angew Phys.*, 1970, **30**, p. 163

14 ROZANOVA, N. B., and GRANOVSKII, V.L.: 'On the initiation of electrical breakdown of a high vacuum gap', *Sov. Phys. Tech. Phys.*, 1956, **1**, p. 471

15 BOUCHARD, K. G.: 'Vacuum breakdown strength of dispersion-strengthened vs. oxygen-free, high-conductivity copper', *J. Vac. Sci. Technol.*, 1970, **7**, p. 358

16 MCCOY, F., COENRAADS, C., and THAYER, M.: 'Some effects of electrode metallurgy and field emission voltage strength in vacuum', Proceedings of 1st international symposium on High voltages in vacuum, Cambridge, MA, USA, 1964

17 KOFOID, M. J.: 'Arc formation between electrodes immersed in a low-pressure plasma', Proceedings of 10th international conference on Phenomena in ionized gases, Oxford, 1971

18 GLEICHAUF, P. H.: 'Electrical breakdown over insulation in high vacuum', *J. App. Phys.*, 1951, **22**, pp. 535–541 and 766–771

Chapter 4

Current interruption in vacuum

4.1 Description of the task

The vast majority of power switching devices spend almost all their service life in the closed position, conducting current to a load. Yet the major effort in their design and development is directed to the relatively infrequent occasions when, on command, they interrupt this current, or to the even rarer occasions, when they operate in anger to clear a short circuit. Both these kinds of event typically last less than a tenth of a second.

When a pair of current carrying contacts separate, be they in vacuum or in a liquid or gaseous ambient, an arc is drawn, which, through its ample conductivity, provides a path for the current to continue. The task of interruption is to change the volume between the contacts from a reasonable conductor to a very good insulator, and do so with great alacrity. The need for haste is evident when one considers the power involved. In a gaseous circuit breaker arc the current density may be in excess of 5 kA/cm^2, and the electric gradient in the arc may be 200V/cm or more [1], thus the power dissipation is of the order of 1 MW/cm^3. No device can withstand this degree of thermal loading for very long.

There is a second reason why restoration of dielectric strength following current interruption must be very rapid, this is the fact that the contact gap must support the high transient recovery voltage (TRV) that the circuit quickly impresses across it [2]. This can be understood by reference to Figure 4.1,

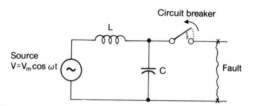

Figure 4.1   Equivalent circuit for studying transient recovery voltage when circuit breaker clears fault [2]

wherein the matter has been reduced to its barest essentials. It is assumed that a load (not shown) is being fed through the circuit breaker and that a short circuit has just occurred, isolating the load from the source. L is all the inductance limiting the current to the point of fault, while C is the natural capacitance of the circuit adjacent to the circuit breaker. It comprises capacitance-to-ground through bushings, current transformers, and so forth, and perhaps the capacitance of a local transformer, as well as capacitance across the breaker contacts. Resistance and any other form of loss has been neglected.

The fault current, being inductive, lags the voltage by 90°, thus when interruption occurs at current zero, the supply voltage is at its peak. While the circuit breaker is arcing, the voltage across its contacts is the arc voltage, which for vacuum is very low. This constraint is removed once arcing ceases, allowing current from the source to flow into C to bring it to source potential. Being a resonant circuit, the voltage of C, and therefore across the switch, overshoots; in short, an oscillation occurs at the natural frequency of the circuit, the period being $T = 2\pi(LC)^{1/2}$. This is the TRV referred to previously; it is illustrated in Figure 4.2.

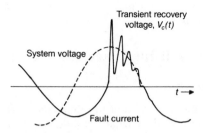

Figure 4.2   Transient recovery voltage across circuit breaker in Figure 4.1 following interruption of fault current [2]

Twice peak voltage is impressed across the switch contacts after $T/2$ seconds. If either L or C is small, or more so if both are small, this peak could be reached very quickly. For example, if the natural frequency is 20 kHz, the peak would be attained in $25\mu s$.

To maintain the electrical conductivity required to carry a short circuit current, the core temperature of the arc in a gas (or oil) circuit breaker must be of the order of $20,000°K$ [3]. It maintains this very high temperature by the energy it draws from the power system; the instantaneous power is the product of the arc current and the arc voltage. In quasisteady state, this intake just balances the losses due to conduction, convection and radiation from the arc. To destroy the conductivity the arc must be cooled, but this is virtually impossible when the current is high, since as heat is extracted, by a blast of cold gas for example, the arc exercises its aforementioned will to survive and increases its arc voltage to compensate for the greater loss. Applied as current zero is approached, cooling can be more successful in its purpose, and at the instant of current zero there is momentarily no power input, which allows the destruction of the energy balance to be completed, the arc extinguished and the circuit isolated.

The process of current interruption in vacuum is different, it involves the rapid dispersal of the arc products, that is the removal of the plasma which had previously been conducting the current in the arc. Chapter 2 presented a picture of the vacuum arc, so we have a good idea of the nature of the beast we are trying to tame. Chapter 3 discussed vacuum breakdown; having interrupted the current, this will help understand how to keep it interrupted.

A figure of merit for any power interrupting device is the product

$$\xi = \left[\frac{dI}{dt}\right]_1 \times \left[\frac{dV}{dt}\right]_2 \qquad (4.1.1)$$

$[dI/dt]_1$ is the rate of decline of current immediately prior to current zero, $[dV/dt]_2$ is the rate of rise of voltage across the open contacts immediately after current interruption. The higher the value of this product, the better the interrupting device. In this regard, vacuum interrupters are far superior to other types, orders of magnitude better in most instances.

Section 2.2 described two modes of the vacuum arc, the diffuse, and the constricted modes. For reasons that will become clear, the constricted arc recovers much more slowly than the diffuse arc, implying that the $[dV/dt]_2$ capability is much lower in that mode. Consider these two modes separately, starting first with the diffuse mode.

4.2   Interrupting a diffuse vacuum arc

4.2.1   *The declining current*

If successful interruption of a vacuum arc depends on the dispersal of the arc products, it behoves us to consider what are those arc products and how do they disperse. We recognise a number of components: metal vapour, metal ions, electrons, gas molecules, metal droplets and/or particles. The model of the arc presented in Section 2.6 has these species being continuously generated and being continuously dispersed. Electrons and metal vapour issue from the cathode surface at the cathode spots. The vast majority of vapour is ionised in the ionisation region (Figure 2.8) and the metal ions so produced flow to cathode and anode as described, where they recombine with electrons and become metal atoms once again. A small fraction escape to the shields where they are similarly removed from circulation.

The cathode spots move, so they leave behind the trails on the contact surface which continue to emit metal vapour until they cool. Cooling is fairly rapid, depending on the thermal diffusivity of the contact material, and of course, evapouration itself is a powerful cooling process in that each atom evapourated takes with it its latent heat of vaporisation. Vapour is similarly produced from the regions immediately surrounding the active cathode spots, where the temperature is high enough for vapour production, but where conditions of temperature and electric field are inadequate to cause electron emission.

Daalder [4] has reported the presence of droplets or particles ejected from the cathode. These are quite small (in the range 1 to 100 m). This has been graphically confirmed more recently by Gellert *et al.* [5]. Such particles can also give rise to metal vapour as they traverse the gap.

Vapour from whatever source disperses because of the high particle gradients. Unlike the ions it is not directly influenced by the electric field although it may exchange energy and momentum with ions when collisions occur. Like the ions, the vapour condenses on cool surfaces it encounters and is thereby removed from the gap.

In the context of interruption, one is particularly concerned with the population in the contact gap and other electrically stressed regions of the interrupter at current zero and immediately afterwards. This requires that one knows what the production rates of the different particles are and how long they stay in the gap, for these are the factors that determine their density or concentration. Production, of both vapour and ions, is dependent on the erosion rate from the cathode. A number of people [4,6,7,8] have measured this and determined that for a diffuse arc on copper it is in the range 50–100 μg/C. Some of this is in the form of droplets. Higher vapour pressure materials have greater erosion, but I concentrate on copper since it is a major constituent in most power interrupter contacts.

A number of investigators, Tanberg [9], Reece [10], Easton *et al.* [11] and Plyutto *et al.* [12], have measured the velocity of the jet of copper vapour emitted from the cathode spot. They used essentially the same method which was to insert a vane into the vapour and measure the force on the vane or the reaction force on the cathode. This allows one to calculate the momentum exchange and knowing the mass of vapour deposited, the mean velocity can be determined. For copper the figure was found to be approximately 10^4 m/s. This means that on the average, vapour particles remain in a 1 cm contact gap for about 1 μs, from which one can infer that at the time of arc extinction the gap has very little memory of prior events.

This observation clearly has important implications for current interruption at power frequency (50 or 60 Hz), where the decline of current zero, $[dI/dt]_1$ in Equation 4.1, is extremely slow on the time scale of vapour dispersal. As Farrall [13] puts it, '. . . a vacuum interrupter will begin to recover while the arc is still burning just after the sinusoidal peak'. How different this is from a gas blast interrupter. In vacuum the power frequency arc has difficulty maintaining itself as current zero is approached. It usually becomes unstable and is interrupted prematurely. This phenomenon is discussed in Section 4.5.

Knowing the vapour velocity and the erosion rate, and assuming a velocity distribution (usually Maxwellian) it is possible to compute the vapour density in the gap at any instant during the decline of current. We are particularly interested, of course, in conditions at current zero, when dielectric recovery proper begins. The basic premise is that early in the recovery period, the interelectrode volume contains a high density of metal vapour which, when high voltage is applied, breaks down through collisional effects in the manner discussed in Section 3.2. As the decay of vapour proceeds, the density of neutrals ultimately approaches a level for which the electron mean free path in that vapour is of the order of the gap length. This condition is taken to be a critical one since at that time, breakdown is assumed to have become independent of the presence of decaying vapour.

An implicit assumption has been made in the foregoing, namely, that when vapour particles reach a solid surface they condense and are removed from consideration in the gas phase. This is frequently the case. However, if the surface is hot, the *accommodation coefficient* is less than unity, that is to say some fraction of the particles rebound from the hot surface, causing the vapour density to be higher than computed. Zalucki and Kutzner [14] have investigated the consequences of this. The effect is more important for prolonged arcing since the flux of vapour itself heats the condensing surfaces. It is particularly important with a constricted arc as Section 4.3 will show.

Now turn attention to the ions in the arc as the current declines and see what effect, if any, they may have on recovery. The model used is that described by Childs and Greenwood [15], and later by Childs, Greenwood and Sullivan [16]. As the current commences to decline there are typically a number of cathode spots pouring plasma into the interelectrode gap. As the current falls, these will extinguish one-by-one until only one remains. Most of the current is carried by electrons but some fraction is carried by ions; for copper this fraction is approximately 8% [17]. In Section 2.6.2, it was suggested that the electrons and ions were like two trains on parallel tracks with the electron train going $(2-s)/(1-s)$ times faster than the ion train according to Equation 2.6.6.

As the current in the last cathode spot continues its decline following the dictate of the external circuit, the electron train must decelerate so that, at current zero, it is travelling at the same speed as the ion train. The gap remains bridged by low impedance plasma, so current continues to flow. The ions have considerable inertia and therefore maintain their progress towards the anode. The electrons, on the other hand, continue to decelerate, or in terms of the train analogy, the electron train goes slower than the ion train and the net current is negative. What we are observing is *post-arc current*.

In a very short time the electron train comes to a halt and to maintain the dI/dt it reverses. However, in doing so, it creates a region adjacent to the anode which is depleted of electrons. It is at this instant when the electrons reverse and the positive ion sheath forms, that the TRV commences to build up and concentrate across the ion sheath.

This scenario leads to the expectation that there should be an observable pause between current zero crossing and the build-up of the TRV. This is clearly evident in Figure 4.3, where the lower trace shows the current coming down to zero, passing through, and displaying a pronounced post-arc current. The upper curve shows the voltage across the switch (Figure 4.5 shows this pause even more clearly). The circuit which provided Figure 4.3 is shown in Figure 4.4 [16]. A current of a few thousand amperes is established in the vacuum switch (VS) by closing T1, which discharges the capacitor C_1. When the voltage on this capacitor attempts to reverse, the diode conducts, changing the circuit from an *LRC* circuit to an *LR* circuit in which the current decays exponentially. Opening the contacts when the current is flowing causes a diffuse vacuum arc to form. T2 is closed at a preselected contact separation, causing C_2 to discharge a reverse current through the vacuum interrupter, thereby bringing the arc current quickly to zero. T1 and T2 are ignitrons.

The rate of change of current at this time, $[dI/dt]_1$ in Equation 4.1.1, is given by V_2/L_2 where V_2 is the instantaneous value of V_2. If current commutation from VS into C_2 is achieved with little change in V_2,

$$\left[\frac{dI}{dt}\right]_1 \simeq \frac{I_{ss}}{t_r} \tag{4.2.1}$$

where I_{ss} is the current at the instant of commutation and t_r is the commutation time, as depicted in Figure 4.5.

In accordance with previous discussion, the current continues more or less with this slope until the electrons reverse, which occurs at point (a) in Figure 4.5.

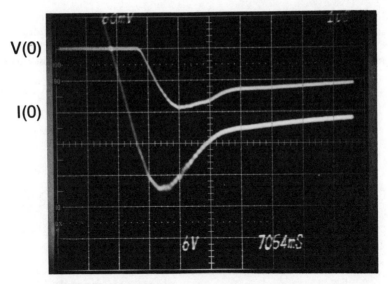

Figure 4.3   *Interval of current zero passage at interruption of diffuse vacuum arc, showing brief pause between current zero and appearance of TRV*

Lower trace current; upper trace voltage
Scale: 1 major div. = 6 kV = 150 A = 1 μs

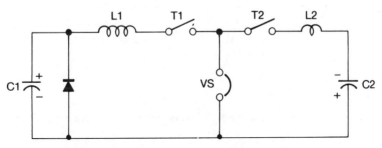

Figure 4.4   *Test circuit for fast commutation of vacuum arc [16]*

It is at this instant we see the recovery voltage buildup since current, entrained in L_1 and no longer having a path through VS, is diverted into C_2. In Figure 4.3, interruption was successful, VS recovered.

4.2.2   Current zero and the immediate post-arc period

If a vacuum switch fails to interrupt a current, that is to say, if a reignition occurs, it means that at least one cathode spot has been established on the former anode. One can only speculate on exactly how this comes about. The end of the last section described how, soon after current zero, a sheath containing positive ion space charge commences to form adjacent to the old anode as electrons are swept from the intercontact gap. There is probably metal vapour in this region

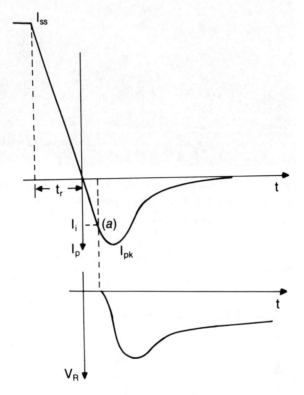

Figure 4.5   Post-arc current and TRV following commutation of vacuum arc [16]

too and there can be considerable electric field, depending on the TRV that develops. Figures 4.3 and 4.5 describe the brief interval of time following current zero. Under the influence of the electric field the ions will be accelerated towards, and will bombard, the former anode, perhaps raising its temperature at local asperities to a point where electron emission can occur. Again, it seems possible that electrons released in this way may create an avalanche in the sheath and that breakdown across the sheath, i.e. between the retreating plasma boundary and the former anode, may result.

Glinkowski [18] and Greenwood and Glinkowski [19] have constructed a model for this period, to examine how the sheath develops, what field is produced at the former anode and what the energy input to that electrode might be. Their work is an extension of earlier work by Varey and Sander [20], Allen and Andrews [21], and by Andrews and Varey [22].

The analysis starts at the instant the sheath begins to form and the TRV starts to develop across it. An important feature of this work is that it takes into

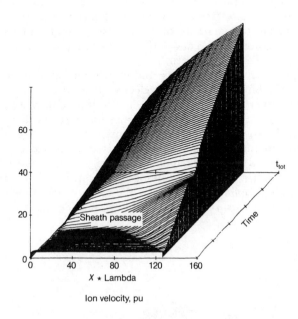

Figure 4.6   Calculated ion velocity distribution [19]

Tungsten
$TRV = 1$ *MHz, 8 kVp*
Initial velocity distribution: uniform, 2.5×10^3 *ms*
Initial density distribution: uniform, 1.5×10^{18} *l/m³*
Boundary condition for velocity: decaying to zero in approx. 23 ns
Boundary condition for density: decaying to ~0.5% in approx. 29 ns
$n \ (base) = 0.8 \times 10^{19} \ m^{-3}$
$t_{tot} = 290$ *ns*
$V_{snd} = 1.02 \times 10^3$
$\lambda = 3.72 \times 10^{-6}$
time s $= 3.63 \times 10^{-10}$
space s $= 2.60 \times 10^{-5}$
Gap length $= 0.5$ *mm*
$T_e = 2 \ eV$
$T_i = 0.272 \ eV$
$E \ (base) = 0.537 \ kV/mm$
$j \ (base) = 1.31 \ kA/m^2$
$v \ (base) = v_{snd}$

(By permission of IEEE)

account the distribution of ion velocity and density at this time. This was difficult because no literature existed on the subject. An attempt was therefore made to bracket the conditions by posing three possible scenarios based on the quasineutrality of the plasma. The approach is similar to Ecker's [23] existence diagrams. The analysis is two dimensional, one dimension in space, one in time, although it is known that the arc plasma flows radially from the cathode spot towards the anode.

The governing equations to be solved are conservation of mass for compressible flow, conservation of momentum, Poisson's equation, and current continuity. Some illustrative results of the analysis for representative conditions are presented in Figures 4.6–4.8 and 4.10. The dimensions of the figures are per unit but conversion factors are given.

Figure 4.6 portrays the ion velocity and clearly indicates the passage of the sheath edge by the line separating the velocity plane with rather low values in the left-bottom corner from the mountain-like surface of increasing velocity on the right, where the increasing field accelerates the particles. For this example the input cathode boundary condition is represented as a decaying function along the left-hand time axis. This simulation terminates after 290ns (not all time steps are plotted) by which time the velocity at the anode has reached approximately 70 km/s. The corresponding density in Figure 4.7 drops rapidly because of its close coupling to velocity by the conservation of mass.

Two separate processes contribute to the plasma decay. First, the sheath

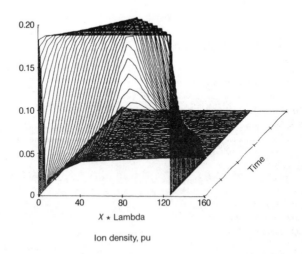

Ion density, pu

Figure 4.7   Ion density distribution [19]

See Figure 4.6 caption for parameters

(By permission of IEEE)

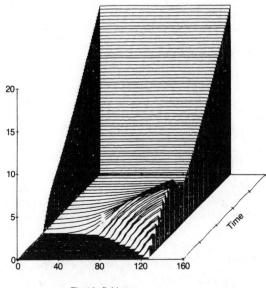

Electric field, pu

Figure 4.8   Electric field distribution [19]
See Figure 4.6 caption for parameters
(By permission of IEEE)

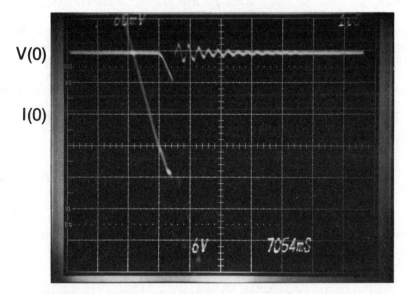

Figure 4.9   Reignition following fast commutation of vacuum arc
Lower trace current; upper trace voltage
Scale: 1 major div. = 6 kV = 150 A = 1 μ s

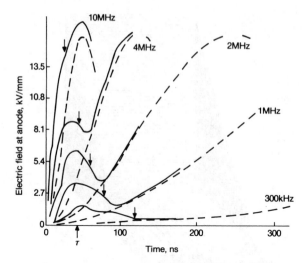

Figure 4.10   Electric field at anode against time for different TRV frequencies

τ = *ion flight time*
——— *effect of space charge distortion*
- - - - *inherent electric field*

'consumes' the plasma as it accelerates ions from the anode side. Secondly, residual plasma diminishes by diffusion, causing density to decline on the cathode side. Note the hump in the electric plot on the anode (right-hand) side of Figure 4.8, peaking at about 3.5 kV/mm at 80 ns. This can be explained qualitatively as follows. When the TRV rises very rapidly because of its high frequency and/or high amplitude, the sheath moves into the plasma before decay on the cathode side can significantly affect its density. This quickly leads to a thin sheath of high density ions, causing a steep electric field distribution and a high E at the anode surface. The condition disappears as the previously mentioned density-reducing processes take effect. Soon thereafter the electric field penetrates the entire gap and becomes more uniform, approaching the value V/d.

If, as suggested, a reignition occurs, precipitating the formation of a cathode spot on the former anode, the early peak in E may be a contributing factor.

The oscillogram in Figure 4.9 was taken from the test which immediately followed the test for Figure 4.3. On this occasion a reignition did indeed occur: the voltage collapsed and the current went off scale. The small steps in Figure 4.8 are caused by the discretising of the variables. They can be eliminated by reducing the space and time steps, without upsetting the stability of the solution, but at a cost of longer computation.

A great advantage of computer modelling is the ability it provides to test the sensitivity of a process or series of events to any of the parameters involved by varying the parameter while keeping the others constant. This has been done in Figure 4.10 for the TRV frequency in the range 300 k to 10 MHz as it affects the electric field at the anode. The dashed lines show the undistorted field; the solid lines show the effect of the plasma. Arrows indicate when the sheath reaches the cathode (note: we are dealing with short gaps). The effect of the moving sheath on the E profile is most pronounced for the 2 MHz TRV. It is almost suppressed at 10 MHz by the overall high E. At the low-frequency end the E peak is low and broad.

One thing that is very evident from Figure 4.9 and the supporting model which gave Figure 4.10 is the severity of conditions required to bring about a reignition, in terms of the $[dI/dt]_1$ and $[dV/dt]_2$ in Equation 4.1.1. We are witnessing a current declining at 540 A/μs, and a TRV climbing at 12 kV/μs. The current slope corresponds to a peak of approximately 1.4 MA at 60 Hz, which is not a commonplace occurrence, to say the least. Nor would the arc be diffuse at this current. The rate-of-rise of the TRV is also exceptionally high for medium-voltage power systems. One must therefore conclude that reignitions during the period immediately following a power-frequency current zero are very unlikely to occur if the arc is diffuse. This accords with the observed facts and supports the contention that vacuum has extraordinary recovery characteristics compared with other switching technologies.

A number of people have made measurements of the rate of recovery of dielectric strength following the interruption of a diffuse vacuum arc; Farrall's [13] experiments are a good example. His approach was to produce an arc across parting contacts with an AC power source, to interrupt the current at (or near) a sinusoidal zero of current, and then electronically switch the AC source off coincidentally with the extinction of the arc. A second voltage source, which was either a simple DC supply or a short pulse a few microseconds wide, was then applied across the gap. The advantage of this arrangement is that the application of voltage from the second source can be delayed from the extinction of the arc by any chosen time interval. By making a large number of successive experimental trials and varying the delay interval for each one, Farrall studied the time dependent properties of the decaying arc column.

The rapidity of dielectric recovery following arcing in vacuum has already been commented on, and also that the power–frequency scale recovery begins before arcing is over. In low power AC circuits vacuum arc recovery is ofttimes so rapid that the decay of arc residue cannot easily be studied. The problem stems from the slow decline in current of a 60 Hz waveform near a sinusoidal zero and the fact that the residue remaining after extinction is very small. This difficulty can be overcome by forcing the current rapidly to zero just after the sinusoidal peak using an auxiliary circuit, much as is done in Figure 4.4. Rich and Farrall [24] have described such a circuit. The time decay of relatively dense residue can then be studied by probing the gap with high-voltage pulses at various time intervals after arc extinction. The magnitude of the voltage required to breakdown the gap is a measure of the degree to which the gap has recovered.

Figure 4.11 gives recovery strength data [13] for a few different electrode metals. These results were obtained using the method just described and show the breakdown voltage of a vacuum gap as a function of time after the forced extinction of arc current. The experimental conditions for each electrode metal were the same.

Obviously the characteristics of the recovery are determined by the properties of the electrode metal. While a small atomic mass is probably significant in improving the rate of recovery, mass is not the sole determinant. Other characteristics such as the initial density of vapour at arc extinction are also

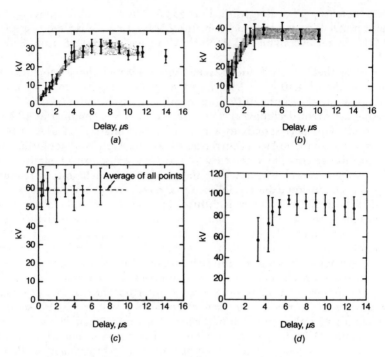

Figure 4.11   Recovery of dielectric strength after vacuum arc [13]

(a) *Silver electrodes*
(b) *Copper electrodes*
(c) *Beryllium electrodes*
(d) *Steel electrodes*

Diameter = 2 in
Gap length = 2.3 mm
Arc current = 250 A (chopped)

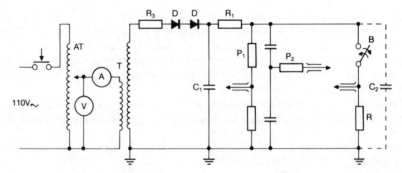

Figure 4.12   Roguski's circuit [25] for obtaining cold gap dielectric recovery characteristic of vacuum circuit breaker

(*By permission of IEEE*)

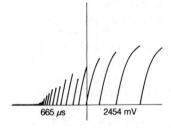

Figure 4.13   Typical oscillogram from cold gap recovery test [25]

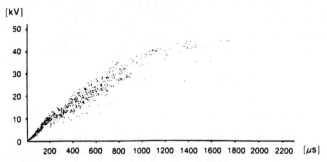

Figure 4.14   Composite data for many cold gap tests [25]

important. Working in the author's laboratory, Roguski [25] examined the dielectric recovery of a vacuum breaker for so-called 'cold gap' conditions, i.e. with essentially no prior arcing. The test circuit he used is shown in Figure 4.12. C_1 is a large capacitor $(3.75\mu F)$ which can be charged to 45 kV by the supply on the left. To perform a test, the test breaker B makes a fast close/open operation. R_1 is such that C_1 discharges very little during the time the breaker remains closed; values used were in the range 100–500 kΩ. When the breaker opened, it immediately interrupted the small current. Thereafter, C_2, which represents the self capacitance of the test breaker and stray capacitance of leads, measuring circuit, etc. is charged through R_1, the time constant being R_1C_2. The small, but increasing gap is overstressed thereby and breaks down, causing C_2 to discharge through the contacts. The relatively small high frequency discharge current is quickly interrupted and C_2 again is charged. This cycle repeats itself many times with, in general, an increase in breakdown voltage. In this way, the successive points of reignition trace out the cold gap dielectric recovery characteristic of the breaker. A typical oscillogram from a test, which illustrates this point, is shown in Figure 4.13. Obviously, the results obtained depend on the speed of opening of the switch.

Collecting together the data from a large number of such tests, it is possible to construct a composite characteristic which conveys a good sense of the recovery rate for the contact material involved. Such a plot (after Roguski [25]) is to be found in Figure 4.14.

It is quite apparent that the data extends well beyond the first few microseconds after current interruption, so that we can expect reignitions and restrikes to occur in practice from time to time, sometimes milliseconds, sometimes cycles after current zero. It is almost certain that such events are initiated by particles or perhaps by flashovers across internal surfaces. Both these kinds of events are discussed at the end of Chapter 3.

Very late breakdowns have recently acquired the name of nonsustained disruptive discharges; their 'discovery' has caused considerable consternation. This is understandable if one's acquaintance with switchgear is confined to other technologies, since a sparkover between the contacts of an open oil or gas-blast circuit breaker would surely lead to the violent destruction of the equipment. In the 1950s we called these (unexplained at that time) late breakdowns 'odd balls', or simply OBs (Hugh Ross once confided that his name for such an event was a 'barnacle'). Whatever, when such a breakdown takes place, it frequently results in a brief high frequency spark as local capacitances share charge. The switch quickly interrupts this current. At worst, there will be a halfcycle of power-frequency current before the circuit is cleared. In most instances this event goes undetected on the system. Further comment on OBs will be found in Sections 11.3.

4.2.3   Post-arc current

Another feature of interest in Figures 4.3 and 4.5 is the *post-arc current*, i.e. the current, seemingly of opposite polarity, which flows after current zero. As explained, the initial part of the current arises from positive ion current and decelerating electrons in the residual plasma. Once the sheath starts to form and the TRV appears (after time t_r in Figure 4.5), the post-arc current has two components, the first due to the scavenging of the remaining ions and electrons from the gap; this is a true conduction current. The second component is a displacement current due to the changing electric flux in the gap, or put another way, the charging of the gap capacitance. Express this as dQ/dt or as $d(CV)/dt$. The C is left within the brackets in the second expression since C as well as V is changing because the gap is increasing. Indeed, the primary capacitance initially is across the sheath because most of the contact gap is shorted out by plasma. When a digital oscilloscope is used to measure the post-arc current and TRV, it is possible to extract the displacement current. Glinkowski [18] has done this; an example is presented in Figure 4.15. The oscillations on the displacement current are not real but computational. The immediate post-arc current (until the TRV begins) mirrors the prior current profile approaching zero: the steeper one is, the steeper the other. Thus, the peak magnitude of the post-arc current depends on dI/dt before current zero.

The post-arc current in the oscillogram of Figure 4.3 is considerable. This is because of the very fast decline of the current. For a typical power-frequency current, say 5 kA at 60 Hz, the dI/dt for the symmetrical wave is only 2.67A/μs. Under the circumstances the post-arc current would be a great deal smaller.

Experiment shows [26,27] that if the conduction portion of the post-arc current is integrated over its duration, it is found that the charge involved is more than the charge remaining in the contact gap at current zero. This strongly suggests that ionisation continues during the post-arc period, but has not reached a point where it causes breakdown.

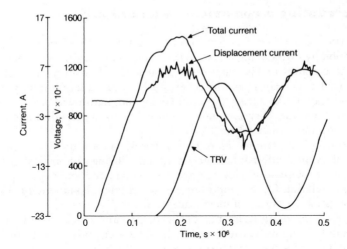

Figure 4.15 *Digital oscillogram of post-arc current and capacitive component of this current, together with TRV*

(By permission of IEEE)

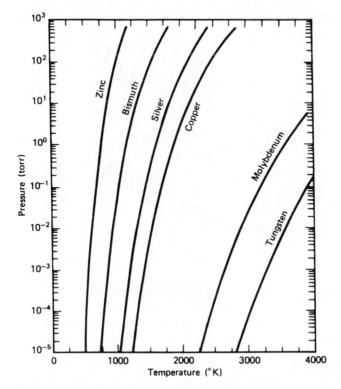

Figure 4.16 *Vapour pressure curves for a number of metals*

4.3   Interrupting a constricted vacuum arc

The anode spot of the constricted vacuum arc dramatically changes conditions for interruption by introducing a new, very copious source of metal vapour. The thermal time constant of the anode spot is such that it continues to produce metal vapour after current zero. Moreover, the hot and vapour-emitting surface of the anode spot is the target for positive ion bombardment once recovery begins. Conditions are therefore more conducive to the establishment of a cathode spot on the former anode.

This is a convenient point to discuss how different contact metals can influence the course of events, interruption or reignition, during the critical recovery period. Properties of greatest relevance are the boiling point and vapour pressure curves of the material. Of less importance is the thermal conductivity. Vapour pressure curves for a number of metals are shown in Figure 4.16; a much fuller tabulation will be found in Reference 28. Lead, zinc and similar metals produce copious amounts if vapour because of their high vapour pressure, thus there is a higher probability of a reignition by ionisation of this vapour when the TRV is impressed across it. At the other extreme, refractory materials like tungsten and molybdenum must attain a very high temperature to produce a sufficient vapour to maintain the vacuum arc because of their very low vapour pressure. The anode spot is thus very hot at current zero and because of the relatively low thermal conductivity has a relatively long thermal time constant. When the recovery period begins refractory metals are still hot enough to emit electrons

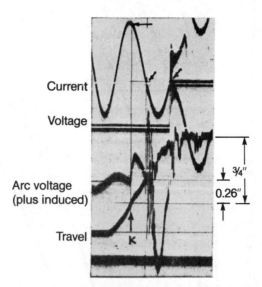

Figure 4.17   Vintage magnetic oscillogram showing a vacuum interrupter reigniting after a major loop of current and clearing at the end of subsequent minor loop

thermionically; the high surface electric field gradient assists in this emission process. Thus, we find that switches made with contacts of such refractory metals reignite and are incapable of interrupting high currents. Materials of this kind are often referred to as 'hard' contact materials in contradistinction to the high vapour pressure materials which are dubbed 'soft'. More will be found on this subject in Section 5.3.

This line of argument suggests that the best contact materials for interrupting constricted vacuum arcs should have vapour pressure characteristics somewhere midway between these extremes, which accords with experimental observations. Such materials are typified by copper. As will be seen, there are a number of quite severe problems which militate against pure copper as a contact material, contact welding is a good example, however, copper is excellent from an interruption point of view and is found as an important constituent in most vacuum interrupter contacts.

Various artifices have been devised to improve the efficacy of contacts for high current interruption. Two good examples are the spiral and contrate contacts illustrated in Figures 1.5 and 1.12. By causing the arc to rotate as a consequence of the self magnetic field of the current, both of these contact arrangements avoid the arc roots dwelling for a protracted period at any one location. In addition, the vapour is spread throughout the volume of the interrupter, so that nowhere is it especially dense when the recovery phase begins.

A second advantage of the rotation is that fluxes of heat and vapour are more uniformly distributed on the shield, thereby helping to assure that no part of the shield becomes so hot that it can no longer act as a condensing surface for the vapour and becomes itself a source of vapour.

As a vacuum interrupter approaches its current interruption limit, it may very well 'skip' a current zero following a major loop of an asymmetrical current and then clear successfully after the following minor loop. Such an event is depicted in Figure 4.17. The peak current was 19 kA; K indicates where the contacts separated, just beyond the peak when the contact gap is 0.26 in. The arc voltage increases as the gap increases. The TRV (second trace) develops at the end of the minor current loop. This is a quite acceptable performance when compared with those of oil and air magnetic breakers which vacuum has displaced. These older devices would sometimes take two cycles or more of arcing before they cleared.

Another extraordinary attribute of vacuum is the abuse the interrupter can take. On a number of occasions when the author was conducting short circuit tests, an interrupter was called on to rupture a current far beyond its rating as a consequence of the inadvertent mis-selection of a reactor tap. The test device failed to interrupt and was isolated by the test station backup breaker. However, when the test was repeated with the correct reactor setting, the interrupter performed impeccably. In like circumstances an oil breaker or an air magnetic breaker would almost certainly have been destroyed.

4.4 Keeping the arc diffuse

4.4.1 *General observations*

The material of the last section suggests a clear incentive for maintaining the vacuum arc in the diffuse state whenever possible. This is a matter of solving a thermal problem. As the current in the arc is raised, the energy spent within the interrupter, which is the integrated product of the current and the arc voltage,

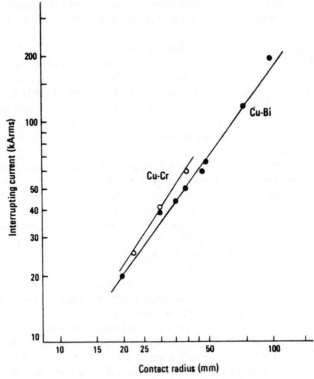

Figure 4.18   Contact radius against interrupting current (rated voltage 6–24 kV) [30]

increases since both elements of the product are increasing. This results in a corresponding increase in the heat flux to the contacts and the shields. In due course this leads to the formation of an anode spot which, though accompanied by a drop in arc voltage, increases the *local* heat flux at the point where the spot forms because of the focusing of the current. At any event, the cat, by this time, is out of the bag. It would therefore appear that to avoid this contingency, one must either increase the area to which the heat flux is applied, or reduce the heat flux itself by reducing the arc voltage. We now examine these two alternatives.

4.4.2   Increasing the internal surface area of the switch

Kimblin [17,29] has shown how the onset of arc constriction can be delayed by increasing the size of the anode; Figure 2.15 illustrates this very clearly. This figure shows that increasing the size of the anode itself reduces the arc voltage. Inasmuch as we are dealing with AC interrupters, it is necessary, of course, to make *both* electrodes essentially identical. Bigger contacts usually mean a bigger enclosure to house them. How the radius of the contacts affects the current interrupting capability is well illustrated in Figure 4.18 which is taken from a paper by Yanabu *et al* [30].

It should come as no surprise that big interrupters are better than small ones for interrupting high currents. However, considerable ingenuity has been shown in the manner in which a large electrode surface area has been designed into a small volume. The arrangements of Rich [31] are a case in point. In most instances the fingers do not move but are attached to the opposing end plates. A pair of central mating contacts carry the current and separate to interrupt. The arc when it is drawn transfers quickly to the fingers as a diffuse discharge. An example of the Rich style of interrupters is shown in Figure 4.19. Such designs have been able to maintain arcs in the diffuse mode up to very high currents. A triggered gap version (Section 10.14) was tested successful through 63 kA at 84 kV [32]. It is my belief that, if required, the interrupting limit could be pushed much higher if one accepted the size of the device required. It is likely that other problems would be encountered in such a design, such as the electrical connections to the switch and containment of the considerable electromagnetic forces generated by the high current, especially the 'popping' forces.

Switches of the finger type have not found commercial favour, probably because of their cost of manufacture engendered by the complexity of the design. However, they could be quite useful for special purposes, such as fusion machine applications.

4.4.3 Use of an axial magnetic field

Section 2.4 noted that the presence of an axial magnetic field (AMF) has a confining effect on the arc plasma which inhibits or delays the formation of anode spots. Much use has been made of this in maintaining the arc in the diffuse mode. Plasma streams radially away from the cathode spot as suggested by Figure 2.13. In the presence of an AMF, the component of ion velocity transverse to the axis creates a mutually perpendicular Lorenz force. Thus, if the velocity component is radial, the force, and therefore the change in trajectory, will be in the circumferential direction. It is this which gives rise to the confining effect. This, in turn, significantly reduces the arc voltage and therefore the power input to the switch. Confirmation of this action was provided by Kimblin's and Voshall's [33] observation that the ion current to the shield diminishes when an AMF is applied.

Lee *et al.* [34] were among the first to report on this phenomenon and suggest that it might be exploited to increase the interrupting capability of devices. The extent of the arc voltage reduction produced by the AMF is evident from Figure 4.20, which is taken from a paper by Morimiya *et al.* [35]. We can observe two effects: first, the arc voltage passes through a minimum as the field is gradually increased, and secondly, for any given field, the arc voltage increases with arc current, which is, of course, the case without an AMF.

As expected, the decrease in arc voltage and power input brought about by the AMF, greatly enhances the current interrupting capability of vacuum switching devices. Some interesting data has been provided by Gebel and Falkenberg [36]. What is remarkable is the magnitude of the arc current, 200 kA RMS, which it is possible to interrupt by maintaining the arc in a diffuse mode and providing large enough contacts. Gebel and Falkenberg [36] provide an interesting diagram (Figure 4.21) to show the arc characteristics at various instants after contact separation. These can be read off along the horizontal axis of Figure 4.21. The time difference from contact separation up to current zero corresponds to the arcing time t_a. Time t is measured from inception of the

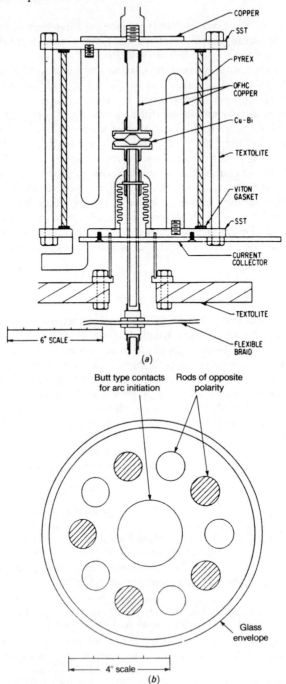

Figure 4.19   Rod-array type of demountable interrupter (after Rich [32])

(a) *Profile*
(b) *Plan view*

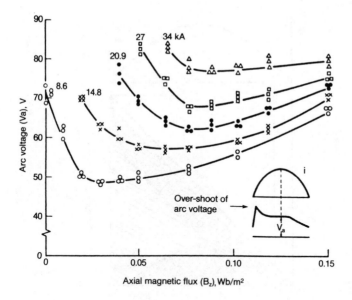

Figure 4.20   Arc voltage against axial magnetic flux for various arc currents; 30 mm electrode spacing [35]

(By permission of IEEE)

current half cycle. One can see, for example, that from approximately 0.5 ms after contact separation, a diffuse discharge spreads over the whole contact surface, while the constricted arc at the point of the last current bridge phases out within 2–3 ms. If the contacts separate 4 ms before current zero (shown by the dotted line), the arc is then evenly diffuse.

With longer arcing times, the discharge spreads diffusely over the whole contact surface and then becomes restricted to the central area of the contacts. This form is, however, quite different from the constricted discharge observed during contact separation. This arc restriction, which results from magnetic forces, is reduced in the second half of the current loop as the current declines and, for example, in the case of an arcing time of 9 ms, has phased out again at least 2 ms before the next current zero as shown by the broken line in Figure 4.21.

Contact structures for providing axial magnetic fields are described in Section 5.3.4.

4.5 Current chopping

4.5.1 *The phenomenon and the implications for power systems*

When a vacuum switching device interrupts a current at power frequency, the current rarely comes smoothly to zero. Before current zero is reached, usually when the instantaneous current is a few amperes, the arc becomes unstable,

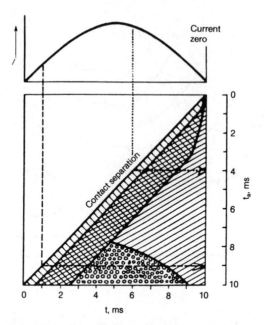

Figure 4.21   Arc-state diagram for contacts with axial magnetic field [36]

    ▨ *Arc constricted after contact separation*
    ▧ *Spreading of diffuse discharge and simultaneous phasing out of constriction*
    ▨ *Diffuse arc*
    ▨ *Arc restriction in centre of contact*

causing the current to be interrupted abruptly and prematurely. Such an event is a *current chop* and the phenomenon is referred to as *current chopping*.

Current chopping is not confined to vacuum interrupters, it has been known for many years that oil and gas blast circuit breakers chop current [37]. However, the cause of chopping in these breakers is instability in the arc column; the overzealous action of the arc quenching devices, designed to control the power arcs of short-circuit currents, succeeds in destroying the conductivity of the slender arc column before current zero when a much lesser current is being interrupted. In contrast, current chopping in vacuum is a contact phenomenon, specifically a cathode phenomenon. Moreover, it is not confined to small currents, but will occur in high currents when the arc is diffuse, witness the chop in Figure 4.22 which the author measured; it occurred after a peak current exceeding 5.5 kA.

Current chopping attracted attention a long time ago because of its potential for creating overvoltages. How this can occur is simply demonstrated in Figure 4.23(*a*) which shows a switch opening to interrupt the magnetising current I_m to a transformer [38]. Figure 4.23(*b*) shows the equivalent circuit and Figure 4.23(*c*) the form of the magnetising current, concluding with the current chop.

Suppose that at the time the chop occurs the instantaneous current is I_o. This is flowing in the transformer winding and is associated with a certain amount of

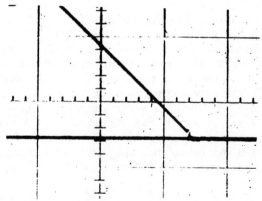

Figure 4.22   *Current chop of few amperes measured by the author following current peak in excess of 5.5 kA*

Scale: 1 div. = 55.2 A = 20 μs

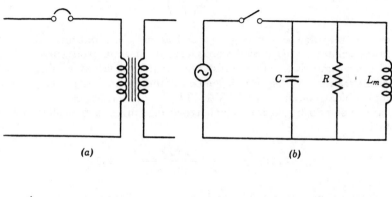

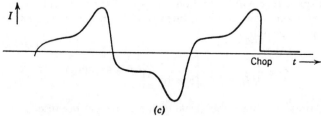

Figure 4.23   *Current chopping [38]*

  (a) Circuit breaker and an unloaded transformer
  (b) Effective equivalent circuit
  (c) Current

  (Courtesy of John Wiley and Sons, Inc.)

magnetic energy, most of which resides in the core:

$$\text{energy} = \tfrac{1}{2} L_m I_0^2$$

This energy may be considerable, for although I_m is only 1% or less of the normal full load current, the magnetising inductance L_m is quite high. The current cannot cease suddenly in such an inductive circuit, yet it has no complete path through the switch after chopping has occurred. It is therefore diverted into the system capacitance on the transformer side of the switch, which is designated C in Figure 4.23(b). This consists primarily of the transformer winding and bushing capacitances, together with any other capacitance that may be in the connections between the switch and the transformer. When the current is diverted in this way, the energy of the magnetic field of the transformer is transferred to the electric field of the capacitance. If this capacitance is known, it is possible to calculate the voltage to which C is charged

$$\tfrac{1}{2}CV^2 = \tfrac{1}{2}L_m I_0^2$$

or

$$V = I_0 \left(\frac{L_m}{C}\right)^{1/2} \tag{4.5.1}$$

This states that the peak voltage reached across the capacitor, and therefore across the winding, is given by the product of the instantaneous current chopped and the surge impedance of the transformer. A striking fact about Equation 4.5.1 is that the transient voltage is independent of the system voltage. To put a figure on this problem, consider a 1000 kVA, 13.8 kV transformer, of the kind found in substations of industrial plants. The magnetising current is typically 1.5 A; thus, at 60 Hz

$$L_m = \frac{V}{\omega I_m} = \frac{13.800}{\sqrt{3} \times 377 \times 1.5} = 14\,\text{H}$$

The effective capacitance will depend on the type of winding and the insulation, whether oil, air or some solid product, but would be in the range 1000–7000 pF. Suppose 5000 pF is chosen, then

$$Z_0 = \left(\frac{L_m}{C}\right)^{1/2} = \left[\frac{14}{5 \times 10^{-9}}\right]^{1/2} = 53,000\,\Omega$$

If the circuit breaker chops the peak current, which because of harmonic distortion might be 2.5 A, the theoretical transient voltage peak would reach 132 kV, which is surely an abnormal overvoltage for a 13.8 kV system. We will have more to say about this in Section 10.5 when dealing with switchgear applications.

4.5.2   *Fundamentals of the chopping phenomenon*

The earliest comprehensive attempt to understand the physical phenomena involved when current is chopped in a vacuum arc was the investigation by Lee and Greenwood [39] in 1961. They devised a mathematical model to describe the mechanism and performed measurements to confirm the model's predictions.

The following four quantities were treated as dependent variables:

J current density in the cathode spot
s fraction of current carried by electrons in the acceleration region
E electric field at the cathode
T temperature of the cathode spot.

In the most general case, all these should be treated as functions of position. For a radial spot they are all functions of the radial distance r from the centre of the cathode spot. However, in the absence of precise information on the manner in which the variables depend on r they were treated as constants over the spot area. For this reason it was convenient to introduce the spot radius a as a fifth dependent variable.

The cathode drop V_c was treated as an independent variable, values being selected from the work of Reece [10]. Other independent variables were the physical constants of the cathode material, the most important of which were

ϕ, the work function
M, the atomic mass
k, the thermal diffusivity and
A, B, C, vapourisation constants [28].

Five equations are needed to determine the five dependent variables uniquely. Lee and Greenwood [39] use the following four:

(i)   Mackeown's [40] space-charge equation relating E, J and a at the cathode
(ii) the T–F emission formulation of Murphy and Good [41] relating T, E, J and s at the cathode
(iii) energy balance at the cathode, considering power input by ion bombardment and taking into account electron emission, evaporation of neutrals, conduction and radiation
(iv) the simple geometric relationship $I = \pi a^2 J$.

Unable to formulate a fifth equation though aware that it '... is undoubtedly intimately concerned with the flow of vapour from the cathode', (a fact confirmed by Harris [42] much later), they introduced two limiting conditions to constrain solutions within narrow bounds.

The first of these was the atom–ion balance, which simply stated that in steady state, or quasisteady state, the ion current density could not exceed the rate of evapouration of atoms from the cathode spot. In fact, according to Harris' model [42], described in Section 2.6.2, the ion flux to the cathode is only half of the ions produced.

The second limiting condition was the magnetohydrodynamic (MHD) equation of motion. The high current density close to the cathode spot creates a $J \times B$ force which tends to pinch the discharge. This is counter balanced by the vapour pressure. For stable operation one would expect the latter to exceed the former for the arc would constrict were this not the case, and this in turn would lead to an instability. Constriction of the arc increases the $J \times B$ force and intensifies the constriction. It transpired that this limit was important for refractory contact materials whereas the atom–ion balance was important for nonrefractories.

The analysis showed that for each cathode material there was a current value, below which a stable arc was no longer possible. This was deemed to be the *chopping level* for that material; it varied significantly from one material to another. Of particular interest are the properties of the metal on which the chopping level depends. The analysis showed that a low chopping level was favoured by a material with high vapour pressure and low thermal conductivity, which accords with Reece's [10] findings.

Such a result should not be surprising with the hindsight of the cathode spot model of Section 2.6.2. A stable vacuum arc requires that the spot be freely emitting electrons and atoms and that these interact in the ionisation zone to produce positive ions. Approximately half of the ions return to the cathode, where they maintain its temperature by their bombardment, and by their presence provide a high space charge field at the cathode surface, and consequently strong emission of both neutral atoms and electrons. As the last cathode spot declines, the emitting area presumably shrinks to preserve the cathode temperature. But this is an inherently unstable situation, for if the temperature drops, electron emission and atom evapouration fall sharply, reducing the ion flux necessary to maintain the temperature; this accelerates the demise of the spot.

A metal with a high vapour pressure can maintain adequate emission to a lower temperature. Lower thermal conductivity helps the spot conserve its heat by reducing the heat flux into the contact as the current declines.

The analytical results have been confirmed many times by experiment [43]. One of the most satisfying experiments that the author has conducted used a vacuum interrupter in which one contact was antimony, a high vapour pressure metal, and the other was molybdenum, a refractory. He found that with the antimony cathode the current chopped was typically 0.5 A, but when the molybdenum contact was the cathode, chops as high as 10 and 12 A were recorded. It is interesting to note that as the test was repeated many times, the apparent chopping level of the molybdenum fell appreciably. This was because antimony from the other more volatile contact began to plate out on the molybdenum. In due course this condensed material began to dominate events when the molybdenum contact was the cathode.

A considerable amount of space has been devoted to this analysis because in my view, it represented one of the earliest attempts at a comprehensive study of vacuum arc phenomena. Also, it led to other analyses. For example, the use of limiting conditions was adopted by Ecker [23,46] in his well-known existence diagrams.

4.5.3   Statistical approach to vacuum arc instability

Farrall [45] took a different approach to the current chopping phenomenon. It had its roots in earlier work on DC mercury arcs by Copeland and Sparing [46], who observed that such arcs once ignited burned for a measurable length of time and then extinguished. They found from a large number of tests that the burning period varied statistically, but the average duration for a given arc current was a reproducible number.

In analysing these data, Copeland [47] considered a large number of independent arcs initiated simultaneously, rather than a large number of

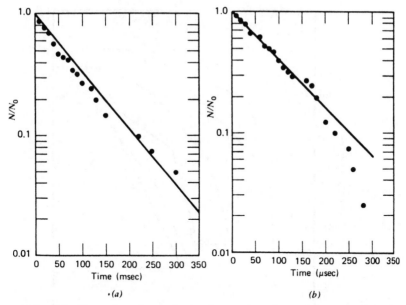

Figure 4.24 Relative number of arcs lasting longer than time t for copper electrodes in vacuum [48]

(a) 3.42 A
(b) 38.5 A

successive ones. He reasoned that the number of arcs dN, which extinguish in time dt, should be proportional to the product of the number of arcs burning at a given instant and the time interval, which is to say

$$dN = -\lambda N dt$$

whence

$$N = N_0 \epsilon^-$$ (4.5.2)

when N_0 is the total number of trials and λ the reciprocal of the average arc duration. This relationship describes very well the data obtained by Copeland and Sparing for given experimental conditions.

Cobine and Farrall [48] performed similar experiments with solid contacts in vacuum and found that the results could be described by the same 'survival' law. Figure 4.24 shows the statistical variation in arc duration for copper electrodes in vacuum at currents of 3.42 and 38.5 A. The results are well represented by Equation 4.5.2. They also observed the average arc duration is greatly dependent on the cathode material as Figure 4.24 indicates.

Since arc stability is strongly dependent on current, Farrall [47] argued that the burning of a power frequency arc should be dominated by the time variation of current near a sinusoidal current zero. To illustrate the consequences of this condition on AC interruption, he imagined that each instantaneous current in the last quartercycle imparts to the arc a stability corresponding to the DC arc

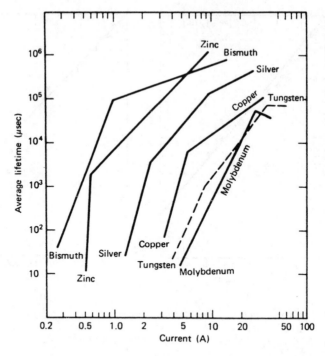

Figure 4.25   Arc lifetimes for different contact materials at different current levels [48]

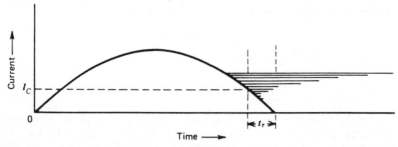

Figure 4.26   Schematic representation of instability at end of AC halfcycle using the DC arc lifetime analogy [45]

(Courtesy of John Wiley and Sons, Inc.)

lifetime for that current. As the current declines, the instantaneous stability also declines. Figure 4.26 schematically represents this 'instantaneous' stability as Farrall conceived it. It shows the instantaneous stability at various points for a declining current as a sequence of horizontal lines that become progressively shorter as the sinusoidal zero is approached. Ultimately, a point in the halfcycle is reached at which the expected 'instantaneous' duration of the arc becomes just equal to the time remaining in the halfcycle. Since the average lifetime is strongly dependent on current, further declines in the current will cause the arc to extinguish shortly thereafter. That is, the current will be chopped. When this deduction is taken together with the data of Figure 4.3.2, it is again clear that low chopping level is favoured by switches with high vapour pressure, low thermal conductivity cathodes.

4.6 High frequency current interruption

4.6.1 Significance in the power switching context

The interruption of high frequency currents is interesting in its own right. Section 1.4 noted that one of the earliest applications of vacuum as a switching ambient was in high frequency current circuits. However, the concern here is with what may be best described as inadvertent interruption of high-frequency currents in the course of switching operations on power systems.

When a switch or circuit breaker closes, the circuit is almost always completed before the contacts mechanically touch. This is because the system voltage, impressed across the diminishing contact gap, creates an increasing electrical stress between the contacts, which results in breakdown before the contacts meet. Such an event is referred to as a *prestrike*. The prestrike current will contain a power-frequency component determined by the instantaneous source voltage and the power-frequency impedance of the circuit. In addition, there will be one or more high-frequency currents that arise as a consequence of capacitances on either side of the switch sharing their charge in local oscillations.

A similar event can occur if the switch reignites or restrikes in the course of opening, for this can be viewed as an inadvertent closing operation. There is an important difference from what has just been discussed, for in this case the contact gap is increasing rather than decreasing. A classical example of such an event is the occurrence of a reignition or restrike when disconnecting a capacitor bank. The magnitude of the current on these occasions can be quite high because the capacitance involved is the capacitor bank itself. For the same reason the frequency of the current is relatively low [38]. Such events should not occur, but they do from time to time. As described in Section 10.6, serious overvoltages may result.

There is the potential for hazardous overvoltages when similar events occur in the course of interrupting the current to a reactor, or breaking the starting current of a stalled motor. The latter situation has been the cause of some concern and the number of investigations in the course of the past several years [49,50,51). How this comes about is now described.

The simplest circuit one can conceive for such an event is depicted in Figure 4.27, which shows a load, represented by L_2, being supplied by a source of inductance L_1; the capacitance represents the natural capacitance of the system including cable, for example, if that is present. Suppose that the vacuum switching device is in the process of opening to isolate the load from the supply. Further suppose that its contacts have separated close to current zero, so that when the switch interrupts only a very small contact gap exists. As the TRV rises across the separating contacts, it is quite possible that a reignition will occur. Should this be the case, current will flow into the gap from both the source and the capacitance C as indicated in Figure 4.27(b) by

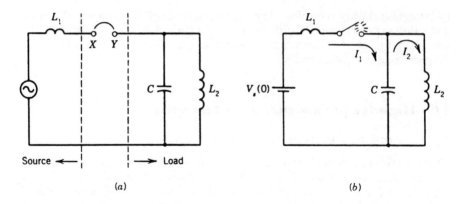

Figure 4.27   Reignition when isolating an inductive load [38]

(a) *Simple representation of the circuit*
(b) *Reignition currents*

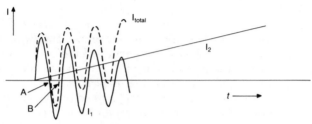

Figure 4.28   Components of the reignition current

currents I_1 and I_2. In Figure 4.28 these currents appear as a function of time, together with their sum, which is the current flowing through the switch. Note that I_1 is oscillatory since the circuit is an LC circuit. Also, the switch current passes through zero at points A and B in Figure 4.28. Because of the extraordinary interrupting capability of the vacuum switch – its very high value of ξ in Equation 4.1.1 – it may well interrupt at one of these instances. If this occurs, a significant overvoltage may be generated. The manner in which this transient voltage is developed is now described.

At instants A and B both I_1 and I_2 are finite, they just happen to be equal and opposite. It is apparent, therefore, that inductive energy is stored in L_1 and L_2 at these times. In the absence of a path for current through the switch, this energy must be delivered to C in the post current zero transient. Unacceptably high voltages can arise as a consequence, which may put the system insulation at risk.

The similarity of these events with what occurs during current chopping caused the author to describe the phenomenon as 'virtual current chopping' a long time ago [52]. Since that time this designation appears to be applied more narrowly to a situation in which the transient current produced by a reignition of a pole in one phase of a three-phase circuit creates a current zero and causes current interruption in another pole.

More is said about the overvoltages produced by reignitions and clearings in Section 10.5.3 where applications are discussed.

4.6.2 Physical phenomena associated with high frequency current interruption

A number of analytical and experimental studies have been made of high-frequency current interruption at short gap length because of the potential consequences just described. The focus has been on the physical events in the gap, rather than the circuit aspects, although it is recognised that the circuit and the decaying plasma in the switch interact. Figures 4.6 to 4.8 and 4.10 were obtained from one such analytical study by Glinkowski and Greenwood [19]. A concurrent experimental study [53] revealed some interesting measured data. Attention was directed to short gaps, since this is the practical condition likely to precipitate a reignition, leading to the high-frequency currents and therefore to the trapping of energy in the manner described.

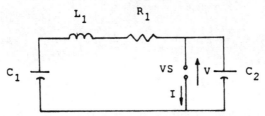

Figure 4.29   Experimental circuit for fixed-gap tests [55]

$L_1 = 5$–$95\ \mu H$; $C_1 = 0.011$–$14F$; $R_1 = 0$–15Ω; $C_2 = 0$–$380pF$

The circuit used for this work is depicted in Figure 4.29 and the vacuum switch under test VS has its contacts *open* at a short fixed gap; the gap selected was in the range 100–150 μm. The capacitor C_1 is charged until the gap of VS sparks over, at which time C_1 delivers an oscillatory current to the gap. The fixed gap assures a reasonably constant breakdown voltage ($\sim$15 kV). dI/dt is controlled by the choice of L_1, whereas changing C_1 changes the frequency and magnitude of the current without affecting dI/dt. Neglecting damping,

$$I_p = V_{C_1}(0)\left[\frac{C_1}{L_1}\right]^{1/2} ; \quad f_0 = \frac{1}{2[L_1 C_1]^{1/2}} ; \quad \left[\frac{dI}{dt}\right]_0 = \frac{V_{C_1}(0)}{L_1} \qquad (4.6.1)$$

The TRV frequency is adjusted by $C_2(\ll C_1)$ as it resonates with L_1. The ranges of values chosen during the course of the tests were

$$50A/\mu s < dI/dt < 1250A/\mu s; \ 100\,kHz < TRV\ frequency < 5\,MHz.$$

A digital oscilloscope was used to obtain a record of the current and the TRV; it was triggered in the first loop of current. A film speed of 20,000 was necessary to provide an acceptable result. Figure 4.30 shows an oscillogram of a test in which there was only one halfcycle of high-frequency current. The upper curve shows the current coming to zero from a negative direction (C_1 was charged negatively); the lower trace is the TRV. Note the pause between the instant of current zero ($t=0$) and the beginning of the TRV ($t=t_0$). This accords with

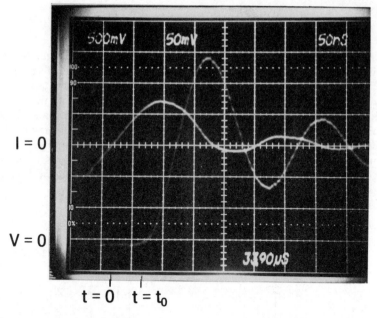

t = 0   t = t₀

Figure 4.30   *Events around high-frequency current zero*
            Scale: 1 div. = 50 ns = 10 A = 2 kV

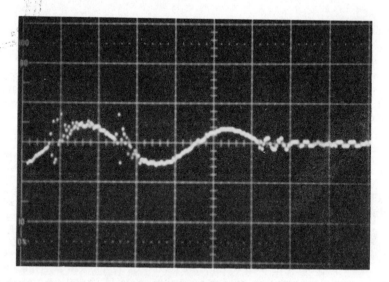

Figure 4.31   *Reignitions while interrupting high-frequency current*
            Scale: 1 div. = 600 ns = 400 A

the explanation of Section 4.2.1; it is a consequence of the sheath not forming at the anode until the electron component of the current is zero. There is still an ion

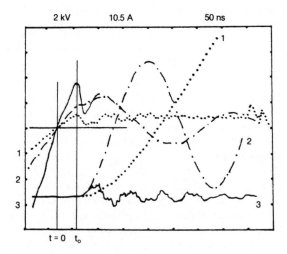

Figure 4.32   Effects of different dI/dt at current zero [53]

    (1) $dI/dt = 161A/\mu s$; $I_0 = 5.5A$; $t_0 = 38ns$; $I_{peak} = 374A$
    (2) $dI/dt = 264A/\mu s$; $I_0 = 9.6A$; $t_0 = 37.3ns$; $I_{peak} = 374A$
    (3) $dI/dt = 674A/\mu s$; $I_0 = 18A$; $t_0 = 40.4ns$; $I_{peak} = 414A$

    (*By permission of IEEE*)

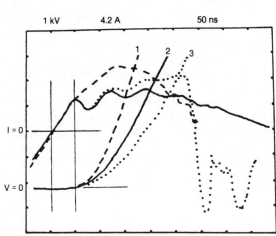

Figure 4.33   Effects of different dV/dt values after $t = t_0$ [53]

    (1) $dI/dt = 121A/\mu s$; $I_0 = 4.1A$; $t_0 = 33.1ns$; $I_{peak} = 374\ kA$; $f_{TRV} = 3.0\ MHz$
    (2) $dI/dt = 124A/\mu s$; $I_0 = 5.3A$; $t_0 = 42.8ns$; $I_{peak} = 3.84\ kA$; $f_{TRV} = 1.51\ MHz$
    (3) $dI/dt = 124A/\mu s$; $I_0 = 4.54$; $t_0 = 39.1ns$; $I_{peak} = 3.84\ kA$; $f_{TRV} = 1.2\ MHz$

    (*By permission of IEEE*)

current flowing towards the anode at this moment with a value of approximately 11A in Figure 4.30. There are again two components of current after t_0:

(i) ion current, just described, plus current produced by ionisation of the residual metal vapour still present in the gap
(ii) displacement current.

This second component, which is proportional to dE/dt in the sheath is in quadrature with the TRV. Figure 4.30 suggests that the conduction current decays monotonically to zero in $\sim$400 ns.

If conditions for interruption are made more severe by increasing $[dI/dt]_1$ and/or $[dV/dt]_2$ (i.e. increasing ξ in eqn. 4.1.1), a point is reached at which the device will begin to skip zeros; it will attempt to interrupt but will reignite. An example is presented in Figure 4.31 which shows the tail end of a sequence and the ultimate clearing. More detailed oscillograms show that the process of forming a cathode spot on the former anode can take some time and be accompanied by current and voltage fluctuations. It is clear, however, that the arc is reestablished after $t = t_0$, indicating that breakdown occurs across the sheath.

Three oscillograms with coincident current zero are superimposed in Figure 4.32 to illustrate how variations in dI/dt affect the interruption process [53]. The pause between $t = 0$ and $t = t_0$ is $\sim$40 ns and is almost constant. This was even found to be true for $dI/dt \sim 1235$A/μs. Note the reignition for trace 3. The value of the ion current at $t = t_0$, when the TRV begins, is therefore proportional to dI/dt, confirming theoretical predictions [16].

Figure 4.33 is again a combination of three oscillograms [53]; in this case dI/dt has been kept constant at $\sim$123A/μs, but the TRV has been varied by changing C_2 in Figure 4.3.2. Reducing dV/dt reduces the displacement current, making it easier to see the conduction current. It shows that an ion current of $\sim$6A flows for about 100 ns after $t = t_0$. There are random variations in this current, from test to test, as it decays. Once more we note a reignition in trace 3.

The ion flight time and therefore the memory of a short gap is of the order of tens of nanoseconds. However, two test series with similar dI/dts but different I_Ps reveal that reignitions are more likely when I_P is higher. The evidence is in the three oscillograms of Figure 4.34 [53]. For test 1, the reignition frequency 77.6 kHz and $I_P = 213$A, giving $dI/dt = 105$A/μs. The corresponding figures for tests 2 and 3 are 4.4 kHz, $I_P = 2150$A, and $dI/dt \simeq 60$A/μs, yet the gap cleared in test 1 but reignited in tests 2 and 3. This suggests that other agents besides dI/dt are at work. As mentioned already, the effect may be thermal; both electrodes may continue to emit vapour, the one from decaying cathode spots, the other from ion bombardment, which is subsequently ionised. Periods of unstable current and voltage precede the breakdowns in tests 2 and 3, indicated by arrows in Figure 4.34.

From this series of experiments one can conclude that when a diffuse vacuum arc current passes through zero, their is a distinct pause before the TRV builds up (approx. 40ns for copper). During this pause the gap carries conduction current only, with an ion component which depends on dI/dt, varying between 3A for 60A/μs and 60A for 1235A/μs. The ion current subsequently decays in tens or hundreds of nanoseconds. It can be distinguished from the displacement current at this time by varying dV/dt, keeping other parameters constant. Among the interruption criteria for short high-frequency vacuum arcs, dI/dt prior to

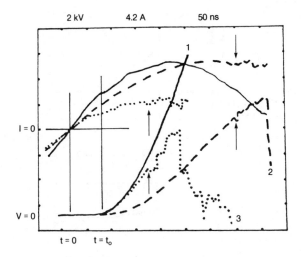

Figure 4.34   Effect of current peak on reignition [53]

(1) $dI/dt = 105A/\mu s$; $I_0 = 5.4A$; $t_0 = 60.2ns$; $I_{peak} = 213\ A$
(2) $dI/dt = 63A/\mu s$; $I_0 = 4.1A$; $t_0 = 66.3ns$; $I_{peak} = 2150\ A$
(3) $dI/dt = 59A/\mu s$; $I_0 = 2.24$; $t_0 = 40.4ns$; $I_{peak} = 2150\ A$

(By permission of IEEE)

current zero and initial dV/dt are the most important. High values of dI/dt are more likely to precipitate reignitions, but breakdowns can occur after lower dI/dts if the gap has been subjected to a high current for a relatively long time ($> 100\mu s$).

The material of the last several paragraphs relates to an investigation with which the author was involved, but similar experiments have been conducted elsewhere, notably by Lindmayer and Wilkening [54].

Figures 4.35(a) and (b), taken from their work, show reignitions following the interruption of high frequency currents. Figure 4.35(b) shows in detail an expanded time scale of one of the reignitions in Figure 4.35(a). It is their belief, as a consequence of the statistical distribution of reignition voltages and as a result of streak photographs that they took, that there are two different reignition mechanisms, a breakdown similar to the cold gap behaviour, and a breakdown at lower voltages initiated in a space charge region in front of the new cathode due to the residual charge carriers. Their streak photography also indicates that reignitions at short gap are localised and repeat time and time again at the same location.

Heyn *et al.* [55] investigated the effect of contact material on the extinction of high-frequency vacuum arcs by adding zinc, antimony and lithium oxide to the more standard 75:25 copper/chromium material. They observed that additives, being more volatile, reduced the reignition voltage and increased the tendency to reignite.

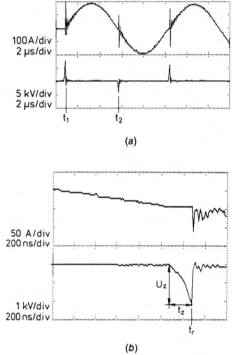

Figure 4.35   (*a*) *Current and voltage oscillograms during high-frequency discharge*
(*b*) *Time-expanded trace*

(*After Lindmayer and Wilkening [54]*)

Paulus [56], and Ecker and Paulus [57] have also investigated short vacuum arcs. Using commercial vacuum devices, they found that short arcs in the range of several micrometres behave differently from longer arcs commonly treated in the literature. By applying *E*–diagrams they concluded that the short arc cathode spot can operate at lower current densities than that of the long arc. They associate this different behaviour with a strong anode influence on the cathode spot. First, the distance from the cathode spot to the anode is small so that anode vapour contributes significantly to the total vapour input and production of arc plasma. Secondly, energy production in the plasma outside the space charge region is no longer negligible. More energy is released in the plasma ball yielding less losses and therefore more stable burning of the arc.

4.7 Interruption of direct current

The traditional way of interrupting a direct current is to have the interrupting device develop an arc voltage in excess of the supply voltage [58,59]. In this way the arc voltage, being opposed to the direction of current flow, drives the current to zero. Thus, if the arc voltage exceeds the supply voltage by ΔV, the rate of

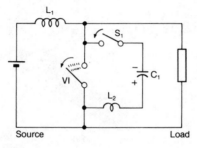

Figure 4.36 Forced commutation DC breaker

decline of current, neglecting circuit resistance, is

$$\frac{dI}{dt} = \frac{\Delta V}{L} \qquad (4.7.1)$$

If it is assumed that $dI/dt = 10^6 \text{A/s}$ and $L = 5\text{mH}$, $\Delta V = 5$ kV. Knowing what we do about the arc voltage of vacuum arcs, the idea of generating 5 kV of arc voltage seems preposterous. It is indeed a fact that the vacuum interrupter is a complete mismatch for breaking direct current by the traditional means.

Direct-current interruption can be effected by forced commutation; a circuit for this purpose is shown in Figure 4.36, which has many of the characteristics of Figure 4.4. Note that it requires a precharged capacitor C [59]. When a fault occurs, the vacuum interrupter VI opens. Shortly thereafter, the 'switch' S1 is closed, thereby initiating a high-frequency counter current through VI. Put another way, the completing of the local circuit through C and L_2, causes the fault current to be commutated from the VI into the capacitor C. The vacuum interrupter has little difficulty in interrupting the current even if the commutation process is quite rapid. This method of DC interruption takes advantage of the demonstrated capability of vacuum switches to interrupt high-frequency currents.

During the commutation procedure, the current in L_1 will change very little, it therefore follows that there will be energy $\frac{1}{2} L_1 I_0^2$ stored in the inductance L_1 at the time of the commutation (I_0 is the current at that instant) and that this will transfer to C since there is no longer a path for current through VI. It is as if the commutating process had chopped a current I_0. Neglecting damping, and in accordance with Equation 4.5.1, we can expect that the voltage of C will be changed by an amount

$$\Delta V = I_0 \sqrt{\frac{L_1}{C}} \qquad (4.7.2)$$

It may well be that this is excessive. If this is the case provision must be made to absorb much of this energy in a surge diverter placed in parallel with C.

More will be found on the applications of this forced commutation technique in Section 10.12.

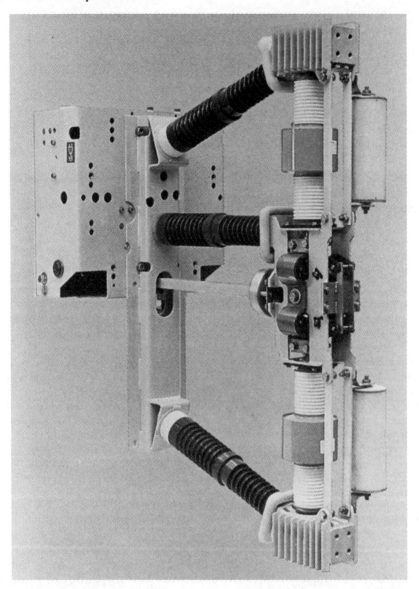

Figure 4.37   Circuit breaker with two interrupters in series, rated 27 kV, 2500 A carrying and 25 kA interrupting

(*Courtesy of Siemens AG*)

4.8   Vacuum interrupters in series

The vast majority of *vacuum circuit breakers* have only one interrupter per pole. High-voltage application require more than one interrupter in series, but such arrangements have difficulty competing economically with SF_6 circuit breakers

Figure 4.38   Load-break switch for 138 kV, 1200A interrupting, with four series interrupters

(Courtesy of Joslyn Hi-Voltage Corp.)

which can achieve the same current and voltage rating with one interrupting chamber. In the move to higher voltages the main thrust is still to lengthen the interrupter to achieve a greater external flashover and use a longer contact gap. Vacuum breakers with two interrupters in series are available; an example is shown in Figure 4.37.

Series interrupters are commonplace in *vacuum switches* and are used for load switching, reactor switching, capacitance switching and line dropping; Figure 4.38 shows such a device. It is an adaption of a simple air disconnect switch. The copper conductors, shown in the vertical, open position in the figure, carry the current when the switch is in the closed position. Blades at the end of these rods, which are now horizontal, engage with jaws at the base of the vacuum interrupter column. Current is commutated to the vacuum interrupters when the blades disengage from the jaws and swing upwards. Final isolation is obtained when an arcing horn attached to the top of the interrupter column falls away from the moving blade. Under ideal conditions, all the interrupters close and open simultaneously and proceed to share equally the recovery voltage, transient and steady state. However, even when simultaneous opening is achieved, equal voltage sharing does not necessarily follow. The distribution of voltage depends for the most part on the distribution of capacitance. Referring to Figure 4.39, we note that in addition to the series capacitances of the interrupters themselves (C_1 in the figure), there is stray capacitance from fittings to ground (C_2 in the figure). One can expect that C_1 will be the same for each series unit, but C_2 will vary because of the geometry, i.e. the physical distances. If n

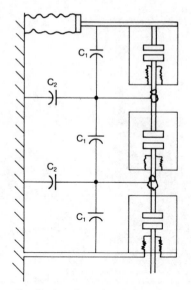

Figure 4.39   Capacitance distribution for series-connected vacuum interrupters

interrupters are in series, and if both C_1 and C_2 are assumed constant, the total shunt capacitance to ground is

$$C_g = (n-1)C_2 \qquad (4.8.1)$$

and for the total series capacitance from end-of-end of the chain

$$C_2 = C_1/n \qquad (4.8.2)$$

Under these circumstances analysis [60] shows that the voltage distribution is

$$V(x) = \frac{V_a \sinh \alpha x/\ell}{\sinh \alpha} \qquad (4.8.3)$$

where V_a is the applied voltage, x/ℓ is the fraction of chain from the ground to end, and $\alpha = [C_g'/C_s]^{1/2}$. The expression in Equation 4.8.3 is independent of frequency, since the network of Figure 4.42(b) contains only one type of element, capacitance. Depending on the value of α, the voltage distribution can have a greater or lesser degree of nonuniformity. With high values of α the interrupters closer to the line end will have a higher proportion of the voltage than those near the ground end.

By way of example, consider a situation in which there are six series units and $C_1 = 100$ pF and $C_2 = 10$ pF. It is apparent that

$$C_g = 50\,\text{pF}, \; C_s = 100/6\,\text{pf}$$

whence

$$\alpha = \left(\frac{50 \times 6}{100}\right)^{1/2} = 1.732$$

The voltage across the interrupter closest to the line is then

$$V_a\left[1 - \frac{\sinh \alpha/6}{\sinh \alpha}\right] = 0.27\, V_a$$

whereas the voltage across the interrupter closest to ground is

$$V_a \frac{\sinh \alpha/6}{\sinh \alpha} = 0.107\, V_a$$

If the distribution were uniform both of these would be $V_a/6 = 0.16\, V_a$.

It is clear that α should be kept as low as possible. Connecting grading capacitors in parallel with each individual interrupter can effect an improvement in this regard, inasmuch as this increase C_1 and therefore C_s. Capacitors for this purpose can be clearly seen in Figure 4.37. It will be seen that there are potential pitfalls to this option.

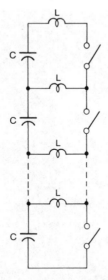

Figure 4.40   Approximate network for series-connected interrupters when considering consequences of reignition

In practice, conditions are less than ideal, so one must expect that the contacts of the series units will not open and close in unison. Sometimes in the course of opening, the contact partings will straddle a current zero. Those contacts which are open will interrupt the current and immediately thereafter will begin sharing

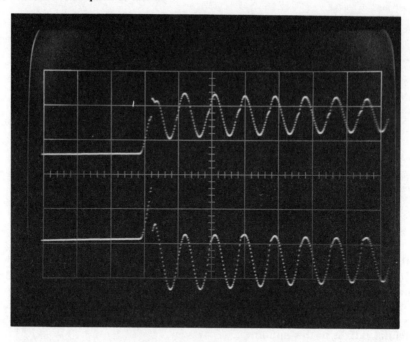

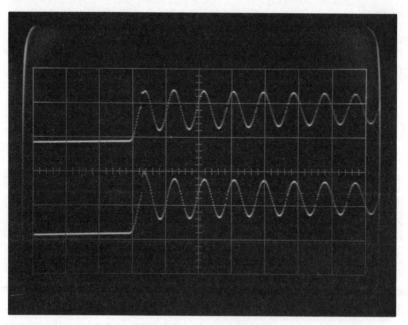

*Figure 4.41   Reignition following current interruption with four series-connected inter-
rupters*

   Traces show how TRVs of four series units are divided between them

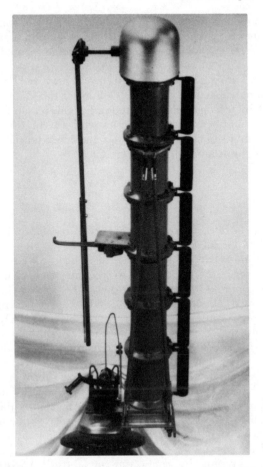

Figure 4.42   *Vacuum switch with six series interrupters with parallel MOVs for voltage grading, designed for 242 kV applications*

(Courtesy of Joslyn Hi-Voltage Corp.)

the TRV. Those contacts which are closed see no voltage across them until they separate. By that time, however, the first group may already have an appreciable voltage across them; the charge associated therewith will remain trapped, so that the different series units will finish up with quite different voltages when the peak of the TRV is reached. This disparity will persist, modified by leakage as time progresses.

Another possibility in these circumstances is that one or more of the highly stressed units will break down. The circuit involved can be crudely represented by Figure 4.40 in which *C* represents the capacitance of each series unit (plus grading capacitor where applied) and *L* the effective inductance between this capacitance and the switch. It is apparent that the discharge current will be multifrequency, with quite high frequency components because of the likely values of the parameters involved. There will be damping of course. It is possible that the delinquent switch will interrupt its reignition current at a high frequency current zero, leaving a quite disparate voltage distribution across the interrupters.

The chance of clearing a high-frequency current may well be enhanced by the addition of grading capacitors since they will lower the frequency of reignition currents (though increasing their magnitude). For this reason it is a good idea to put noninductive snubber resistors in series with any grading capacitors that may be used, adding sufficient, but not too much resistance, to overdamp the circuit. The surge impedance of a local circuit is

$$z_0 = \left(\frac{L}{C}\right)^{1/2} \text{ ohm} \tag{4.8.4}$$

where L is likely to be a fraction of a microhenry. A resistor somewhat in excess of $2z_0$ should be used for the snubber [60].

When a reignition occurs in the manner described, it is entirely possible that another interrupter may be caused to reignite and change the voltage distribution yet again. On several occasions the author has observed the voltage being tossed about like a hot potato between a number of series interrupters. The *total* voltage showed no sign of this, it was only apparent when the individual voltages were recorded. An example of this kind of thing is shown in Figure 4.41, which shows the TRVs of four interrupters connected in series. The second interrupter breaks down temporarily and its share of the voltage is assumed by the remaining three. However, it quickly recovers and proceeds to support voltage thereafter. Note how the polarity of the TRV of interrupter # 2 is frequently opposite to those of the other three! An event of this kind can leave some or all units highly stressed. One must be concerned with *external* flashover in these circumstances. Some designs safeguard against this by encapsulating the interrupters in foam.

Forced voltage grading by connecting metal-oxide varistors (MOVs) in parallel with the interrupters, has proved very successful; Figure 4.42 shows an example. The use of the MOVs permits six series interrupters rather than eight to be used for this 242 kV switch. It is appropriate to compare this switch with the 138 kV device in Figure 4.38, which has no MOV grading. Switches of this kind are suitable for load switching, or line dropping when the charging current does not exceed 100A. They cannot be used for capacitor bank switching because the energy absorbing capability of the MOVs is inadequate.

4.9   References

1 SWANSON, B. W.: 'Current interruption in high-voltage networks' *in* REGALLER, K. (Ed.): 'Theoretical models for the arc in the current-zero regime' (1978) pp. 137–184

2 GREENWOOD, A.: 'Electrical transients in power systems' (Wiley, 1991, 2nd edn.) pp. 50–57

3 REGALLER, K., and REICHERT, K.: 'Introduction and survey' *in* Reference 1, p. 12

4 DAALDER, J. E.: 'Components of cathode erosion in vacuum arcs', *J. Phys. D*, 1976, **9**, p. 2379

5 GELLERT, B., SCHADE, E., and DULLNI, E.: 'Measurement of particles and vapor-density after high-current arcs by laser techniques', Proceedings of 12th international symposium on Discharges and electrical insulation in vacuum, 1986

6 KIMBLIN, C. W.: 'Erosion and ionization in the cathode spot region of vacuum arcs', *J. App. Phys.*, 1973, **44**, pp. 3074–3081

7 RONDEEL, W. G. J.: 'Cathode erosion in the vacuum arc', *J. Phys. D*, 1973, **6**, p. 1705

8 KUTZNER, J., SEIDEL, S., and ZALUCKI, Z.: 'Distribution of neutral particle flux ion the vacuum arc', Proceedings of 9th international symposium on Discharges and electrical insulation in vacuum, 1970

9 TANBERG, R.: 'On the cathode of an arc drawn in vacuum', *Phys. Rev.*, 1930, **35**, p. 1080

10 REECE, M. P.: 'The vacuum switch part 1 – properties of the vacuum arc', *Proc. IEE*, 1963, **110**, pp. 793–802

11 EASTON, E. C., LUCAS, F. B., and CREEDY, F.: 'High velocity streams in vacuum arcs', *Trans. AIEE*, 1934, **53**, p. 454

12 PLYUTTO, A. A., RYZHKOV, V. N., and KAPIN, A. T.: 'High speed plasma stream in vacuum arcs' *Sov. Phys. JETP*, 1965, **20**, p. 328

13 FARRALL, G. A.: 'Recovery of dielectric strength after current interruption in vacuum', *IEEE Trans*, 1978, **PS–6**, pp. 360–369

14 ZALUCKI, Z., and KUTZNER, J.: 'Streams of neutral particles reflected from the anode in vacuum arc', Proceedings of 5th international symposium on Discharges and electrical insulation in vacuum, 1972

15 CHILDS, S. E., and GREENWOOD, A. N.: 'A model for DC interruption in diffuse vacuum arcs', *IEEE Trans.*, 1980, **PS–8**, pp. 289–294

16 CHILDS, S. E., GREENWOOD, A. N., and SULLIVAN, J. S.: 'Events associated with zero current passage duiring the rapid commutation of a vacuum arc', *IEEE Trans.*, 1983, **PS–11**, pp. 181–188

17 KIMBLIN, C. W.: 'Vacuum arc ion currents and electrode phenomena'. *Proc. IEEE*, 1971, **59**, pp. 546–555

18 GLINKOWSKI, M. T.: 'Behavior of vacuum switching devices for short gaps', Doctoral thesis, Rensselaer Polytechnic Institute, Troy, NY, USA, 1989

19 GLINKOWSKI, M., and GREENWOOD, A.: 'Computer simulation of post-arc plasma behavior at short contact separation in vacuum', *IEEE Trans.*, 1989, **PS–17**, pp. 45–50

20 VAREY, R. H., and SANDER, K. T.: 'Dynamic sheath in a mercury plasma', *Brit. J. App. Phys.*, 1965, **2**, pp. 541–550

21 ALLEN, J., and ANDREWS, J. G.: 'A note on ion rarefaction waves', *J. Plasma Phys.*, 1970, **4**, pp. 187–194

22 ANDREWS, J. G., and VAREY, R. H.: 'Sheath growth in a low pressure plasma', *Phys. Fluids*, 1971, **14**, pp. 339–343

23 ECKER, G. *in* LAFFERTY, J. M. (Ed.): 'Vacuum arcs – theory and application' (Wiley, 1980) p. 249

24 RICH, J. A., and FARRALL, G. A.: 'Vacuum arc recovery', *Proc. IEEE*, 1964, **52**, p. 1293

25 ROGUSKI, J.: 'Experimental investigation of the dielectric recovery strength between the separating contacts of vacuum circuit breakers', *IEEE Trans.*, 1989, **PD–4**, pp. 1063–1069

26 GLINKOWSKI, M., and GREENWOOD, A.: 'Some interruption criteria for short high-frequency vacuum arcs', *IEEE Trans.*, 1989, **PS–17**, pp. 741–743

27 GLINKOWSKI, M.: 'Behavior of vacuum switching devices for short gaps', Doctoral thesis, Rensselaer Polytechnic Institute, Troy, NY, USA, 1989

28 DUSHMAN, S. *in* LAFFERTY, J. M. (Ed.): 'Scientific foundations of vacuum technique' (Wiley 1962 2nd edn.) pp. 696–699

29 KIMBLIN, C. W.: 'Anode voltage drop and anode spot formation in DC vacuum arcs', *J. App. Phys.*, 1969, **40**, pp. 1744–1752

30 YANABU, S., TSUTSUMI, I., YOKOKURA, K., and KANEKO, E.: 'Recent tecnical developments of high voltage and high power vacuum circuit breakers',

Proceedings of 13th international symposium on Discharges and electrical insulation in vacuum, Paris, 1988, pp. 131–137

31 RICH, J. A.: 'A means of raising the threshold current for anode spot formation in metal vapor arcs', *Proc. IEEE*, 1971, **59**, pp. 539–545

32 RICH, J. A., FARRALL, G. A., IMAM, I., and SOFIANEK, J. C.: 'Development of a high-power vacuuum interrupter', EPRI report EL-1895, 1981

33 KIMBLIN, C. W., and VOSHALL, R. E.: 'Interruption ability of vacuum interruptèrs subjected to axial magnetic fields', *Proc. IEE*, 1972, **119**, pp. 1754–1758

34 LEE, T. H., KURTZ, D. R., and PORTER, J. W.: 'Vacuum arcs and vacuum circuit interrupters', CIGRE report 121, 1966

35 MORIMYA, O., SOHMA, S., SUGAWARA, T., and MIZUTANI, H.: 'High current vacuum arcs stabilized by axial magnetic fields', *IEEE Trans.*, 1973, **PAS–92**, pp. 1723–1732

36 GEBEL, R., and FALKENBERG, D.: 'Arc behavior on vacuum switch contacts with axial magnetic field', *Siemens Forsch. Entevicklings ber.*, 1987, **16**, pp. 72–75

37 YOUNG, A. F. B.: 'Some researches in current chopping in high voltage circuit breakers', *Proc. IEE*, 1953, **100**, p. 337

38 Reference 2, p. 93

39 LEE T. H., and GREENWOOD, A.: 'Theory of the cathode mechanism in metal vapor arcs', *J. App. Phys.*, 1961, **32**, pp. 916–923

40 MACKEOWN, S. S.: 'Cathode drop of an electric arc', *Phys. Rev.*, 1929, **34**, p. 611

41 MURPHY, E. L., and GOOD, R. N.: 'Thermionic emission, field emission and transition region', *Phys. Rev.*, 1956, **34**, p. 1464

42 HARRIS, L. P.: 'A mathematical model for cathode spot operation', Proceedings of 13th international symposium on Discharges and electrical insulation in Vacuum, Albuquerque, NM, USA, 1978

43 LEE, T. H., GREENWOOD, A., and POLINKO, G.: 'Design of vacuum interrupters to eliminate abnormal overvoltages', *Trans. AIEE*, 1962, **81**, pp. 376

44 ECKER, G.: 'The existance diagram – a useful theoretical tool in applied physics', *Z. Naturforscha*, 1971, **26**, pp. 935–939

45 FARRALL, G. A. *in* LAFFERTY, J. M. (Ed.): 'Vacuum arcs – theory and application' (Wiley, 1980) p. 195

46 COPELAND, P. L., and SPARING, W. H.: 'Stability of low pressure mercury arcs as a function of current', *J. App. Phys.*, 1945, **16**, pp. 302–308

47 COPELAND, P. L.: 'Accuracy of constants in expontential decay as obtained from finite samples – a review', *Am. J. Phys.*, 1945, **13**, pp. 215–222

48 COBINE, J. D., and FARRALL, G. A.: 'Experimental study of arc stability-I', *J. App. Phys.*, 1960, **31**, pp. 2296–2304

49 GUPTA, B. K., DICK, E. P., GREENWOOD, A., KURTZ, M., LAUBER, T. S., LLOYD, B. A., NARANG, A., PILLAI, P. R., and STONE, G. C.: 'The insulation capability of large AC motors, vols. 1 and 2', EPRI report EL-5862, 1988

50 CORNICK, K. J., and THOMPSON, T. R.: 'Steep-fronted switching voltage transients and their distribution in motor windings, part 2', *IEEE Proc. B*, 1982, **129**, pp. 56–63

51 ORACE, H., and MCLAREN, P. G.: 'Surge voltage distribution in line-end coils of induction motors', *IEEE Trans*, 1985, **PAS–104**, pp. 1843–1848

52 GREENWOOD, A.: Discussion of MARANO, M., FUJII, T., NISHIKAWA, H., NISHIKAWI, S., and OKAWA, M.: 'Voltage escalation in interrupting inductive current by vacuum switches' and 'Three-phase simultaneous interruption in interrupting inductive current using vacuum switches', *Trans. IEEE*, 1974, **PAS–93**, p. 278

53 GLINKOWSKI, M., and GREENWOOD, A.: 'Some interruptoin criteria for short high-frequency vacuum arcs', *IEEE Trans.*, 1989, **PD–17**, pp. 741–743

54 LINDMAYER, M., and WILKENING, E.-D.: 'Breakdown of short vacuum gaps after current zero of high frequency arcs', Proceedings of 14th international symposium on Discharges and electrical insulation in vacuum, Santa Fe, NM, 1990, pp. 234–241

55 HEYN, D., LINDMAYER, M., WILKENING, E.-D.: 'Effect of contact material on the extinction of vacuum arcs under line frequency and high frequency conditions',

Proceedings of 36th IEEE-Holm conference on Electrical contacts, Montreal, Canada, 1990
56 PAULUS, T.: 'The short vacuum arc – part I: experimental invetigations', *IEEE Trans.*, 1988, **PS–16**, pp. 342–347
57 ECKER, G., and PAULUS, T.: 'The short vacuum arc – part 2: model and theoretical aspects', *IEEE Trans.*, 1988, **PS–16**, pp. 348–351
58 BOEHNE, E. W., and JANG, M. J.: 'Performance criteria of DC interrupters', *Trans. AIEE*, 1947, **66**, pp. 1172–1180
59 Reference 2, p. 158
60 Reference 2, pp. 327–333
61 Reference 2, pp. 75–77

Chapter 5

Design of vacuum switchgear 1: the vacuum interrupter

5.1 Preamble

There are three principal parts to a piece of vacuum switchgear:

- vacuum interrupters
- mechanism to operate the interrupter
- auxiliary equipment and the packaging which integrate the switching device into the power system.

This first design chapter treats the interrupter; the other parts are dealt with subsequently. How the parts come together, especially the adaptation of the mechanism and the interrupter to the various ratings required by the user, is of particular importance. Accordingly, space is set aside to discuss this matter under the heading of design for versatility.

Electrical, mechanical and thermal considerations enter into the design process, and not least economics, for the product must not only meet the customer's technical needs, it should also be available at an affordable price. We should consider all these aspects in the search for the elusive synergism we call good design.

5.2 Fundamental requirements for the interrupter

Vacuum interrupters are at once very simple and very sophisticated. Simple in that their geometry and appearance is simple, they have few parts; sophisticated, in that much scientific and technical know how goes into the preparation and assembly of those parts. In principle, all that is required is a pair of contacts, a vacuum-tight envelope to enclose them and provide for their support, insulation to isolate the contacts from one another when in the open position, and a shield to maintain the integrity of the insulation by protecting it from the products of the arc - condensing metal vapour - when the switch opens. In as much as the contacts must be separable, the penetration to the moving contact must permit the required movement. This is almost always obtained by a metallic bellows.

Two basic designs have evolved to meet these requirements; they are illustrated in Figures 5.1(a) and (b). In the first style, the ceramic (or glass) envelope, designated 32, provides the end-to-end insulation which is protected by a cylindrical metal shield, designated 33.

The alternate design of Figure 5.1(b) has the shield forming a major part of the vacuum envelope; bushings at both ends insulate the contact shanks. Figures 1.2

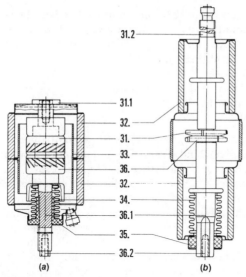

Figure 5.1 Basic designs for vacuum interrupters

(*a*) *Internal shield*
(*b*) *External shield*

(*Courtesy of Siemens*)

and 1.6 are archetypes of the first style, whereas the experimental interrupter shown in Figure 1.7 is an early version of the second. Contemporary vacuum interrupters exhibit many variations to improve performance, reduce cost, or enhance manufacturability, but these are matters of detail, the basic designs of Figure 5.1 are still evident.

The degree of vacuum typically employed is 10^{-6} to 10^{-7} torr. Vacuum interrupters will function satisfactorily at higher pressures, but vacuum of the degree cited is not difficult to achieve and maintain with modern technology. Moreover, as seen in Section 1.2, the vacuum tends to improve as the switch is operated, by the pumping action of the metal vapour and the gettering of the metal when it is freshly condensed.

We now consider in more detail the components described and how they fit together to produce an effective interrupter, starting with the contacts. These are important in any mechanical switching device because they conduct the current in their closed position and support the voltage when the switch is open. However, in the vacuum switch they also provide the arcing ambient in which the transition from the closed to the open position takes place.

5.3 Contacts

5.3.1 Material: a shopping list of ideal requirements

A list of attributes for an ideal contact material for a vacuum interrupter would surely include the following:

- good electrical conductivity to minimise losses when the switch is closed
- good thermal conductivity to conduct away such losses as there may be
- good dielectric properties when the switch is in the open position
- good current interrupting capability
- easily fractured welds
- mechanical stability in the face repeated impact
- good machinability
- easily outgassed

Good electrical conductivity in metals is accompanied by good thermal conductivity (Weidmann and Franz Law) since both heat and electricity are transported through the metallic lattice by electrons. However, some of the other desirable characteristics are in serious conflict. For example, refractory materials such as tungsten provide good dielectric strength and produce brittle welds, which are favourable points, but tungsten has a high chopping level and is a poor high-current interrupter. Its low vapour pressure accounts for the chopping characteristic (Section 4.5) and also for the inadequate interrupting performance. The contacts become exceedingly hot, locally, in the course of arcing and have not cooled sufficiently by the time current zero is reached to preclude thermionic emission from the former anode, now the cathode, in the following recovery period. This leads to reignition of the arc. In addition, tungsten is hard to machine.

Antimony, on the other hand, has a high vapour pressure and therefore produces almost no current chopping, but the profusion of vapour remaining at current zero makes it also a poor interrupter of high currents. The pressure during the immediate post current zero period is too high to support the voltage being applied across it.

It is apparent, as pointed out in Section 4.3, that good short circuit current interrupting capability requires a contact material with an intermediate vapour pressure, not so high that it has to reach a very high temperature to produce vapour, and not so low that the high current arc boils off a vast amount of vapour. Copper suggests itself. The problem with pure, clean copper is that two surfaces simply pushed together in vacuum form an excellent cold weld, which, because of the ductility of copper, is very difficult to break. With the assistance of a prestrike, which causes arcing when the contacts close, conditions for welding are even more favourable.

5.3.2 Solutions to the contact material problem

Lack of good prospects for contact materials among the pure metals led to investigations of alloys and composite material. The result was the development of two materials, quite different from each other, that now account for most of the contacts produced worldwide. The first material, developed in the US [1], is copper–bismuth (Cu/Bi). The second, copper–chromium (Cu/Cr), was originally a British development [2], but was adapted and modified by people in other countries.

A comprehensive discussion of Cu/Bi and some other similar materials is to be found in a paper by Barkan *et al.* [3]. First, it should be stated that the material is principally copper, which is therefore referred to as the primary constituent. The amount of bismuth present, the secondary constituent, is typically 0.5-2%. This means that its characteristics are essentially those of copper, which are for the most part very good, or at least acceptable, with respect to the requirements listed in Section 5.3.1. The purpose of the bismuth is to solve the welding problem, which it does very well.

Among the contributing factors that bring about the desired welding performance are the following:

(i)   Bismuth is substantially soluble in copper *in the liquid phase*.
(ii) Bismuth is almost insoluble in copper in the solid phase.
(iii) The freezing point of Bismuth (271°C) is much lower than the freezing point of copper (1084°C).

Since the two constituents are mutually soluble in the liquid phase, they are readily formed by casting. As the homogeneous liquid cools below the freezing temperature of the primary constituent, it begins to solidify in a granular structure. As the two constituents are essentially mutually insoluble in the solid phase, the liquid becomes increasingly rich in bismuth during this initial cooling period. Finally, when the temperature of the mix has dropped sufficiently, the bismuth solidifies in the grain boundaries of the copper grains. The extensiveness and thickness of the secondary constituent around the primary grain varies according to the relative quantities of the two constituents.

Pure copper is highly ductile. The effect of adding even a small amount of bismuth is to produce a dramatic reduction in the ductility. Cu/Bi is brittle, which is important for the butt contacts of a vacuum interrupter since the energy required to fracture a brittle weld is substantially less than for a ductile material.

The explanation of this phenomena is as follows [3]. The prestriking of closing contacts, the forcing apart by popping forces, or the contact separation due to bounce, all give rise to a momentary arc, which melts the surface locally. The consequence is a shallow molten zone adjoining an essentially cool bulk substrate of contact metal. The two surfaces come together on a molten film which is characteristically only a few microns thick. As soon as contact is made, arcing ceases, the energy input into the contact zone drops to a low value, and rapid cooling begins. The cooling process under these conditions is highly directional since a thin molten film floats on a relatively cool solid mass. Freezing then occurs from the liquid–solid interface in towards the center. The interface between the contacts freezes last. This rapid directional cooling gives rise to two effects shown in Figure 5.2: the grain structure is columnar; and the segregation of the two constituents takes place because of the difference in freezing temperatures. As the copper freezes first, it displaces the still molten bismuth towards the hottest region in the centre of the weld zone. Thus, the last region to freeze is extremely rich in bismuth which freezes in a plane along the interface between the two contacts. The existence of this highly segregated interface in the weld zone apparently accounts for the weak welds and the manner in which they fracture.

The down side of materials of this kind is that the different changes in density of the two constituents as they cool causes significant shrinkage stress to develop as cooling progresses. In the case of thick, continuous films, this results in a slow, but inevitable extrusion process which causes the migration of particles of the secondary constituent to the free surface.

Section 3.3 noted that the presence of weakly bonded particles on the internal

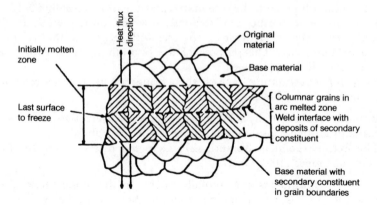

Figure 5.2   Schematic diagram of weld cross-section in copper bismuth [3]

surfaces of a vacuum interrupter can seriously reduce its dielectric integrity. The message, therefore, is to use sufficient bismuth to produce weak welds, but no more than this amount.

Regarding other attributes, 1% Cu/Bi has a conductivity close to 90% of that of copper. Its current chopping level is lower than copper and is entirely acceptable for almost all applications.

Robinson's British patent [2] provides a considerable amount of information regarding Cu/Cr, at least as originally conceived. The chromium powder, comprising particles up to 250 μm in size, is made into a solid block by first compressing it with considerable force and then sintering the fragile green material under high vacuum to produce a mechanically stable chromium sponge. The sintered compact is then impregnated with molten copper, again under high vacuum and at high temperature. The procedure, at least as originally performed, was to place a disc of copper on a disc of the sintered compact, like a pat of butter on a crumpet. When the copper melted it flowed into the interstices of the higher melting point chromium. The result on cooling was a compact mass.

Later in this book, where manufacturing of contacts is discussed, the forming of Cu/Cr matrices is shown to be often done in a hydrogen atmosphere. The hydrogen is readily removed later by heating in vacuum.

The technique originally used tungsten for the sintered material. However, it was found that W/Cu had limited current interrupting capability for reasons already discussed. In spite of the presence of the copper, the anode surface was raised to a high temperature when high current arcs (10 kA) were drawn. This was apparent from observation of significant melting of the tungsten. It was evident that under these conditions the contact surface had reached a temperature at which copious electrons had been generated by thermionic emission. The condition of high electron emission persisted through the instant of current zero into the next halfcycle, causing a reignition of the switch. The lower boiling point of chromium, 2665°C, assured that this did not occur with Cr/Cu, with the desired results with respect to high current interruption.

An impression of the Cr/Cu matrix as depicted in the patent [2] is shown in Figure 5.3. The chromium particles, which are designated 30 and are hatched,

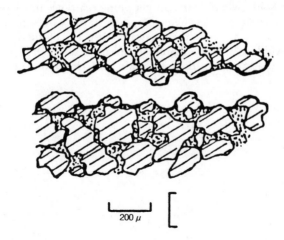

Figure 5.3   Microstructure of chromium/copper contact material

are in contact with their neighbours; the copper 31, shown dotted, fills the interstices. In contradistinction to Cu/Bi, the secondary constituent, in this case copper, typically occupies 10–30% by volume of the impregnated matrix as determined by the porosity of the sintered compact. The chromium powder on compaction yields a matrix metal of low ductility, so that the infiltrated matrix likewise exhibits a low ductility and favours brittle welds.

Over the years different manufacturers have changed the formulation and processing (more is said about this in Chapter 9). The tendency is to have a finer grain structure and to use more copper, to the point where it becomes Cu/Cr rather than Cr/Cu. A 60% copper, 40% chromium is a popular combination, but a higher percentage of chromium gives better performance. A more recent development is the addition of up to 5% vanadium to the Cu/Cr. It is found that this reduces the spread in dielectric breakdown data, thus this material is a candidate for higher voltage interrupters.

Referring again to the matter of fine grain structure, experience shows that this can be brought about by repeated arcing and fast cooling of the Cu/Cr contact surfaces. Such procedure is now introduced into the contact manufacturing process by some manufacturers (Chapter 9).

Other contact materials have been used and are being used, copper–tungsten, for example, is found in some switches where high current interruption is not a requirement. Special materials have been developed to give a low chopping level for use in contactors. Kurasawa *et al.* [4] report on studies with silver-tellurium and silver–selenium, which form eutectics. To reduce cost, a third constituent (cobalt is mentioned) is added to produce a ternary alloy. Copper–lead is a popular material for low-voltage application. The purpose of the lead is to reduce the weld strength. As little as 0.2% is sufficient to accomplish this.

Obtaining high-current interrupting capability and simultaneously a low-current chopping characteristic, presents quite a challenge with respect to contact material. Recent developments indicate that silver–tungsten carbide (Ag/WC) may be a candidate for this application. The solubility of WC in Ag is negligible; the two materials remain separated in the matrix. The current instability threshold, where current chopping can occur, depends on the percentage of the conductive element, the silver in this case. This is indicated in Figure 5.4. The trick is to obtain the maximum dispersal of the Ag by small grain size, thereby avoiding the arc hanging up in a silver region. This is not straightforward inasmuch as the materials with low Ag concentration are

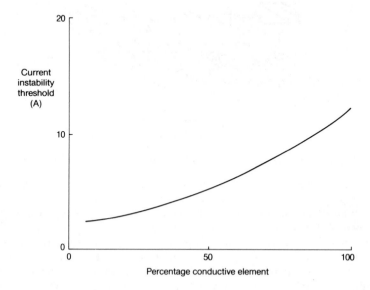

Figure 5.4   Current instability threshold for silver–tungsten carbide

difficult to produce. An acceptable compromise appears to be 25% Ag. This gives satisfactory surge performance and at the same time provides a contact material capable of interrupting 40 kA. However, it is necessary to increase the axial magnetic field by 25% to achieve this result. With this combination the discharge is observed to spread over a larger area of the contacts, especially in the early part of the travel. Kaneko *et al.* [5] give a qualitative explanation of this behaviour.

Slade [6] has provided a rather comprehensive review of contact materials for high-power vacuum interrupters in his keynote address to the 1992 international conference on *Electrical contacts.*

5.3.4 Contact geometry

The size and shape of contacts depends on the job they have to perform. For many applications, where the current to be carried and interrupted is small (a few hundred to a few thousand amperes), simple, cylindrical butt contacts will suffice. For contactors, these may take the form of 'buttons' of contact material, brazed to copper shanks, as illustrated in Figure 5.5.

For higher ratings of current interruption some form of arc rotator is used. The preferred styles are the spiral contact shown in Figure 1.5 and the contrate contact which is featured in Figure 1.12. Interrupters embodying such contacts

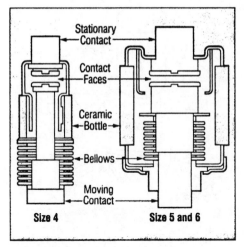

Figure 5.5   Button-type contacts in contactors

(Courtesy of Westinghouse Corp.)

are shown in Figures 5.1(*a*) and (*b*), respectively. Generally speaking, the diameter of the contact, spiral or contrate, is increased as the current interrupting rating is increased. However, as pointed out later in this chapter, advances in design have tended to reduce dimensions.

For ease of fabrication, some manufacturers have simplified the design of the slots in the spiral contact. An example is to be seen in swastika design of Figure 5.16, which is readily produced by end milling the blank.

The advantages of an axial magnetic field (AMF) for keeping the arc diffuse and thereby increasing current interruption was noted in Section 4.4.3. Providing the AMF is a technological challenge. There are basically two ways in which it can be done. The first is to put a coil on the *outside* of the interrupter and excite it by the current through the switch. This gives a fairly uniform field in the region of the contact gap, but considerable eddy currents in the shield of the interrupter will introduce a substantial phase change between the internal field and the external current. Such an arrangement requires a considerable amount of conductor material and presents a problem from a dielectric point of view. A relatively elaborate insulation system is required to ensure that there are no flashovers from the coils to neighbouring structures during service life.

The second method is to make the field producing element a part of the contact structure and utilise the vacuum for insulation. An example of this type is shown in Figure 1.16. This approach also has its drawbacks. The contact structures can become quite complicated and relatively expensive to produce. One must bear in mind that the contacts are under considerable force when in the closed position and the moving contact experiences quite severe dynamic forces during opening and closing. Two examples of contact structures are shown in Figure 5.6. The data in Figure 4.22 were obtained with interrupter incorporating the design of Figure 5.6 (*a*).

Both these sets of contacts display a common feature, namely, the radial slots

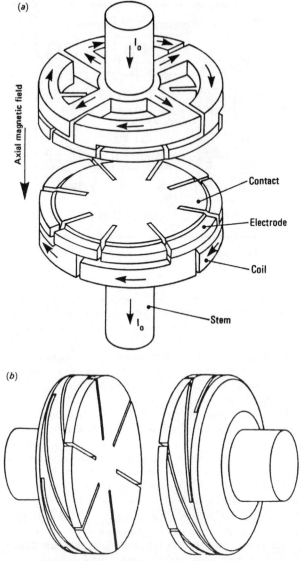

Figure 5.6   Contact structures for providing axial magnetic field

   (a) *After Yanabu et al. [6]*
   (b) *After Benfatto et al. [7]*

in the faces of the contacts. Their purpose is to reduce eddy currents which are induced by the changing magnetic flux of the coil behind. Such eddy currents will tend to cancel the applied field to some extent and also introduce a phase shift between the net field and the arc current. These effects can be quite important in certain local regions. This is born out by the evidence of Figures 5.7 to 5.9. The first of these figures shows the contact structure itself; note that the

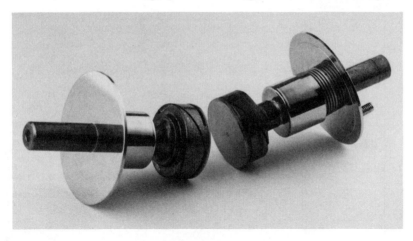

Figure 5.7   Contact structure for vacuum interrupter which provides for axial magnetic field
(Courtesy of Cooper Power Systems)

face of the contact does *not* have slots cut in it. Bestel *et al.* [9] used commercial software to make a two-dimensional plot of the magnetic field. The rotational symmetry of most of the structure makes the two-dimensional approach acceptable. There will be some error due to the lack of rotational symmetry in spiral elements that form the coil.

Figures 5.8(*a*) and (*b*) show the results of the field plots. Figure 5.8(*a*) has a DC excitation, therefore there are no eddy currents.

Figure 5.8(*b*) has AC excitation and the field patterns in the two plots are remarkably different. In Figure 5.8(*a*) the field lines encircle the coil segments in a smooth way, penetrating the other segments almost undisturbed except for a minor change of direction in the end rings that are made of a magnetic steel with relative permeability of 10,000.

In Figure 5.8(*b*) the 60 Hz currents in the source regions induce 60 Hz eddy currents of considerable magnitudes. These prevent the magnetic field from penetrating inside the contact structure. As Figure 5.8(*b*) indicates, contact rings (5) and contact disc (4) are particularly effective in shielding the source currents from the plasma region (6).

Another way of looking at the problem is presented in Figure 5.9 which shows the flux density B_z along the straight line parallel to the contact surfaces, approximately 100 μm above the bottom contact. The suppressing effects of the eddy currents for the AC case is immediately apparent, especially towards the centre of the contact.

The basic structure of the contacts in Figures 5.6(*b*) and 5.7 is the contrate design, first described and illustrated in Chapter 1, Figure 1.12. It is there described as a means for *rotating the arc*. This is because one contact is the mirror image of the others; so that the current produces a *radial* component of magnetic field. The AMF version has a more oblique pitch, but more important, the direction of pitch is the same for the two contacts, so they are like two primitive helices. The currents therefore combine to produce the axial field. The pitch of the helix is a compromise between the requirements

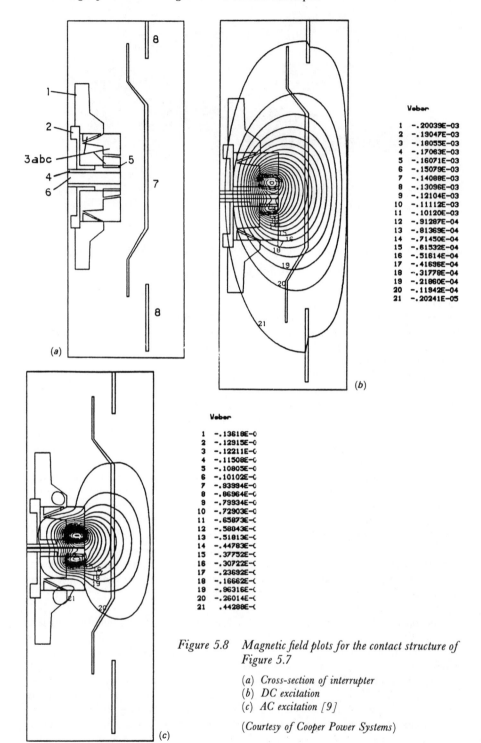

Figure 5.8   *Magnetic field plots for the contact structure of Figure 5.7*

(a) *Cross-section of interrupter*
(b) *DC excitation*
(c) *AC excitation [9]*

(*Courtesy of Cooper Power Systems*)

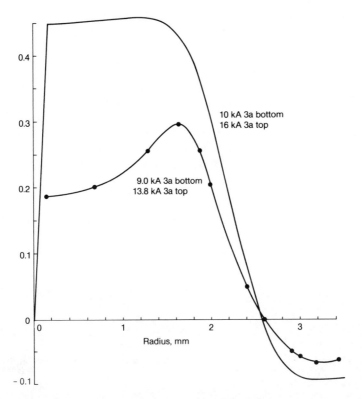

Figure 5.9   Flux density as function of radius for DC and AC excitation of contact structure of Figure 5.7 [8]

(*Courtesy of Cooper Power Systems*)

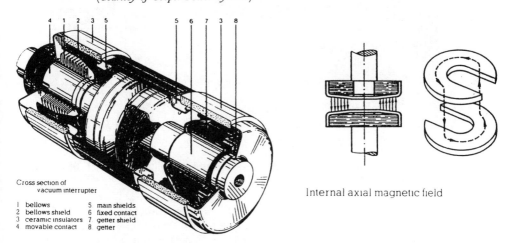

Cross section of
vacuum interrupter

1 bellows   5 main shields
2 bellows shield   6 fixed contact
3 ceramic insulators 7 getter shield
4 movable contact 8 getter

Internal axial magnetic field

Figure 5.10   Producing axial magnetic field by flux-guiding laminations

(*Courtesy of Holec*)

of the magnetic field and the desire to keep the overall resistance low. In a good design the complete interrupter has a resistance of $10\mu\Omega$, which includes the contact resistance.

There are two other intriguing arrangements for producing an axial magnetic field. The first is illustrated in Figure 5.10. Behind each contact in a cylindrical stainless steel box is a set of mild steel or iron laminations, stacked up to form a horseshoe shape, but one in which the legs are chamfered. The openings of the horseshoes at the two contacts point in diametrically opposite directions. The ferrous material tends to concentrate the magnetic field produced by the axial current. In the regions of the horseshoes this flux makes a complete loop, traversing the contact gap in an *axial* direction. When the arc is drawn it tends to concentrate in the regions where the axial field is strongest. When it moves, it moves quickly from one of these regions to the other.

The contacts themselves are disc-shaped, with a rounded, slightly domed surface. They are quite thin and contain no slots.

The flux-guiding elements add considerable weight to the interrupter.

Figure 5.11   Simple axial-field contact structure produced by twisting a cage

(Courtesy of Joslyn Hi-Voltage Corp.)

The second arrangement for producing an axial magnetic field is both simple and effective, Figure 5.11.

The contact is attached to its shank by about eight thick copper wires, which form a kind of cage. The joints are brazed. A short cylinder of ceramic is placed within the cage during fabrication and captured there. The contact is then twisted axially with respect to the shank, causing the copper wires to form a partial helix. The ceramic cylinder prevents the wires collapsing, and gives the contract structure some rigidity.

5.4 Bellows

A metal bellows is almost universally used as a vacuum-tight way of permitting motion of the moving contact. One end of the bellows is fixed, being connected to one end plate of the interrupter, the other moves with the shank of the moving contact to which it is attached (see Figure 5.1). Both seamless and welded bellows are available but the seamless variety are preferred for this application. They are typically made from 347 stainless steel, but other materials such as Inconel X are used. The thickness is 0.1 mm or less.

In most metal bellows applications the motion, and therefore the stress, is applied relatively slowly, causing the bellows to elongate or compress uniformly. In the vacuum interrupter application, however, action is impulsive, one end is moved suddenly through the stroke of the switch and then just as suddenly arrested. It is somewhat like applying an impact to a spring when motion is initiated, and then applying a reverse impact to terminate the motion. In a close–open operation this whole procedure is rapidly repeated, a second time, in reverse order.

The motion imparted to the bellows by such action is no longer uniform, rather a stress wave of rarification or compression, depending on whether it is a closing or opening operation, is caused to travel down the length of the bellows. As with waves in other systems (e.g. the voltage wave when a transmission line is energised) the wave reflects at the switch end, thereby modifying the stress. Indeed, waves travel back and forth along the bellows in a series of reflections until damped out by losses, which in this mechanical system is a relatively slow business. One might suppose that traveling waves among the convolutions would put added stress on the bellows metal as compared with the slow application of force, which is in fact the case. Such stress is cause for concern since it affects the fatigue life of the bellows (the number of operations that can be performed before fatigue cracks cause violation of the vacuum). It is particularly important for contactors which must perform a large number of operations in their lifetime.

Barkan [10] has shown that the behaviour of a bellows under impulsive excitation can be quite adequately analysed by treating the bellows as a uniform rod. Certain characteristics such as the stiffness and the mass per unit length must be known, but these can be obtained from the manufacturer or from simple tests or measurements. The bellows' manufacturer also provides empirical information on the life rating of his bellows when subjected to slowly imposed motions which are free of vibrations. Barkan's paper [10] develops a rather general analytical procedure for determining the dynamic stress conditions in a bellows, subjected to a defined impulsive motion at one end, the other end being fixed.

The rod mode, mentioned above, is analysed to determine the ratio $\Psi = \sigma_m/\sigma_s$ of the maximum dynamic stress range arising in response to the specified dynamic excitation, to the stress which would arise quasistatically if the specified displacement were applied infinitely slowly. The ratio Ψ can then be applied to the stroke-fatigue data provided by the bellows vendor, treating the dynamic overstress as equivalent to an increased stroke to determine a modified fatigue-life prediction.

The analysis shows that the maximum stress always occurs at the moving end and that the maximum dynamic overstress Ψ is a function of the ratio T/τ,

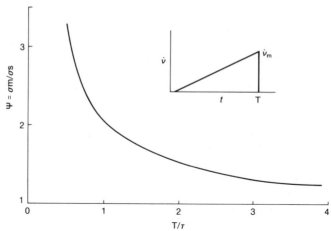

Figure 5.12   *Dynamic overstress factor for uniformly accelerated motion terminated in impact* [10]

where T is the duration of the imposed motion and τ is the time for a wave to propagate through the bellows and back. This last point is illustrated in Figure 5.12. The inset shows the velocity profile of the imposed impact. When the impact is short compared with the travel time, the overstress can be considerable and the corresponding decrease in fatigue life significant. As the stress is applied more slowly, Ψ diminishes, being asymptotic to unity in the limit.

In the course of assembling a vacuum interrupter, the bellows must be brazed to the end shield and contact shank. This subjects the bellows to a high temperature which has an important annealing effect on the material. The properties with respect to fatigue life deteriorate as a consequence. A second objective of the Barkan analysis [10] was to devise a means for estimating the reduction in fatigue life resulting from a complete annealing. With tensile fatigue-life data for both the as-received and fully annealed states of the bellows metal usually available from the manufacturer, he showed how a modified life curve could be developed from the vendor's life-stroke data. The method is graphical; it is illustrated in Figure 5.13.

A combined plot is made of the vendor's life-stroke data for the bellows in the as-received condition and tensile fatigue-life property data for the bellows material. The two plots share the common abscissa of fatigue life in cycles. Two property curves are plotted, one corresponding to the fatigue life of the material in the as-received condition, and a second curve for the metal in the fully annealed state which, it is assumed, corresponds to the condition of the bellows metal after undergoing the high-temperature braze cycle.

With the three curves plotted, it is a simple matter to develop a predicted bellows life curve in the fully annealed state. The arrows and the sequence numbers 1 to 5 illustrate the procedure. For any bellows stroke the vendor's data, curve A, predicts a life corresponding to point 2. In turn, this life corresponds to a tensile strain level defined by point 3 in curve C (incidentally, the 1/2 hard on curve C of Figure 5.11 corresponds to the as-received material). In the annealed state this strain level corresponds to a life defined by intersection 4 on curve D.

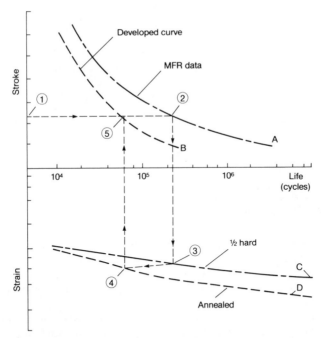

Figure 5.13   Correction procedure for effect of annealing on predicted bellows' life [9]

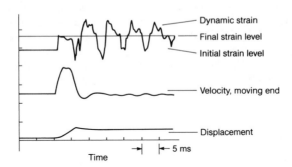

Figure 5.14   Data from a dynamic test on a bellow [10]

Thus, for the given bellows stroke, the annealed bellows should have the same reduced life, and one may, therefore, establish point 5 as one point on the new life curve. The complete life curve B for the annealed bellows can be developed by proceeding in this manner for several values of bellows stroke. It is interesting that annealing typically reduces the life of a bellows by a factor of four.

A good correlation was found between measured and calculated results when bellows were instrumented with strain gauges and the means to accurately determine displacement. Figure 5.14 shows an oscillogram from such a test.

Consistent with the theory, the stress waves are clearly seen to have a shape corresponding with remarkable precision to the imposed velocity. The ratio of

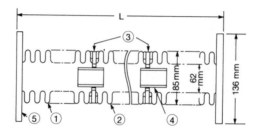

Figure 5.15   Anti-buckling design for long bellows [10]

     1, 2 Bellows
     3 Jointing flange
     4 Buckling prevention pipe
     5 Movable side flange

dynamic stress to static stress is readily obtained from the measurements. It is reassuring that reasonable agreement was found with theory.

The message this work conveys from a design point of view is that it pays to avoid making the bellows too short and to recognise the life-cycle derating that must be applied to accommodate annealing during brazing.

The material just noted acquired a new significance with the advent of designs to increase the operating voltage per interrupter module. Such applications call for a longer stroke to achieve a greater contact separation in the open position and an increased opening speed to compensate for the higher rate of rise of transient recovery voltage. Both these requirements impact the bellows design. As the total length of the bellows is increased, there is a greater tendency of the bellows to buckle, which makes uniform contraction and extension of the convolutions more difficult. Tsutsumi *et al.* [11] suggest a method of combatting this problem by dividing the bellows into a number of sections which are joined by flanges, to which are attached buckling prevention pipes as illustrated in Figure 5.15.

5.5   Shields

As the name suggests, shields are there to protect. The principal shield in Figure 5.1(*a*) (designated 33) prevents vapour from reaching and condensing on the principal insulation of the vacuum envelope (designated 32) where it would ultimately short out the interrupter. The design is different in Figure 5.1(*b*), but the result is the same. The central part of this device, being metal, needs no shield, but note how this is extended axially to obstruct the line of sight from the arcing region at the contacts to the inner insulating surfaces of the bushings. In most designs the centre shield is floating because, as noted by Greenwood *et al.* [13], the switch performance is better that way.

The centre shield serves another function, namely, to relieve the electrical stress on the joints where the ceramic or glass is attached to the metal. These so-called triple points (conductor, insulation and vacuum) can be the source from which discharges leading to breakdown are propagated. Minimising the stress at

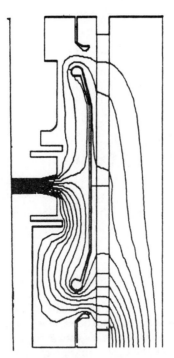

Figure 5.16   *Electric field plot for one-half of an axisymmetric vacuum interrupter showing the influence of the shields [14]*

such locations reduces the chance of a breakdown. This sometimes calls for field plotting in the course of interrupter design. Sokolija *et al.* [14] discuss this matter and present some plots, one of which is reproduced in Figure 5.16. The field plotting was carried out with the aid of a finite element package. Figure 5.16 has the axis of the switch on the left, with the contact gap at the center and the shield and ceramic to the right. The other half of the interrupter would, of course, exhibit the mirror image. Surface conditions of shields are important from a dielectric point of view. It is customary to electropolish or otherwise treat them, as explained in Section 9.2.2.

A shield is also placed over or close to the bellows, its purpose being to protect the bellows from molten debris from the contacts which might conceivably cause their puncture. This shield is often cup-shaped, as in Figure 5.1(*a*) and 5.7, or it may be a simple disc as in Figure 5.1(*b*).

Higher voltage, longer vacuum interrupters may have a number of shields in a relatively elaborate arrangement, as indicated in Figure 1.13.

A variety of materials have been used and are being used for shields. These include stainless steel, nickel, a nickel/cobalt/iron alloy, and copper; the stainless steel may be nickel plated to a thickness of a few microns. In some copper shield designs, where the clearance from the floating shield to the contacts has been reduced to limit the overall diameter of the interrupter, a cylindrical liner of contact material is brazed to the inside of the shield, opposite the contact gap. It is found that if the arc impinges on the shield, it is less likely to hang to this

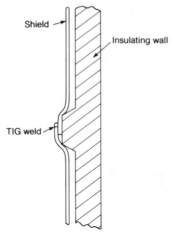

Figure 5.17   TIG-weld joint of two-part shield

material than would be the case with copper[*]. This avoids gross erosion of the shield.

The method of supporting the floating shield is important; it should be rigid. Anything less secure may produce particles when subjected to vibration, and these could adversely affect the dielectric integrity of the interrupter. A popular method of support is to have the central flange brazed between the two halves of the ceramic enclosure. This joint must, of course, be vacuum tight. Other interrupters have the support for the shield cast into the glass or ceramic envelope. The flange of the shield is then tack welded to the support. An alternative is to make the shield in two parts (this saves space) and join them by inert-gas welding (TIG) as illustrated in Figure 5.17. This design tends to be less rigid.

5.6   Vacuum enclosure

In the style of Figure 5.1(*a*), the vacuum envelope comprises principally the end plates and the ceramic or glass insulation between. The centre shield becomes part of the envelope in the style of Figure 5.1(*b*). The most common material for the insulation is high-purity alumina, which may be glazed or unglazed. An example of a ceramic envelope appears in Figure 5.18.

Glass is also used, indeed, all the early interrupters were made of glass since the technologies from which they were developed were based on glass. The glass of the envelope shown in Figure 5.19 is made by a centrifugal casting process which allows the support for the floating shield and the spinnings for attaching the end plates to be sealed into the material as it is formed. The ceramics on the other hand must be metallised at their ends so that they can be brazed to these fittings. More is said about this in Section 9.2.3.

The end plates, to which the contacts are mounted, and which are themselves used to mount the interrupter in the circuit breaker, are typically made of

---

[*] P. O. Wayland, private communication

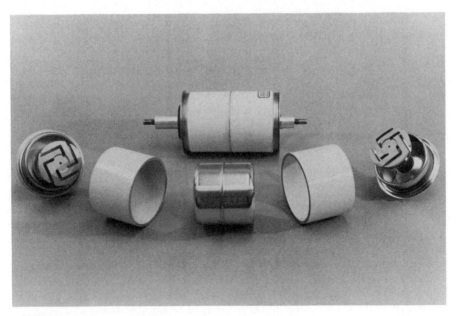

Figure 5.18   Principal parts of a vacuum interrupter
(Courtesy of Westinghouse Corp.)

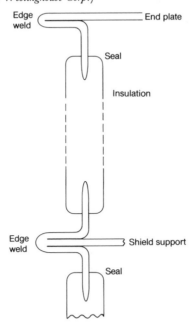

Figure 5.19   Spinnings for end plates and shield support sealed into centrifugally cast glass

stainless steel. They are relatively thin (1–3 mm) to provide some flexibility. Some thinner end plates have radial ribs pressed in them for added strength.

Sealing the vacuum enclosure can be done in two ways, the details are presented in Section 9.2.4. The first takes the subassemblies - the subassemblies of one interrupter are set out in Figure 5.18 - and brazes them together, forming brazed joints at both ends (ceramic to end plates) and at the centre (ceramics to floating shield). The procedure is carried out under high vacuum, that is to say in a vacuum furnace. In the second method the subassemblies are welded rather than brazed, in the manner shown in Figure 5.19. It is a vacuum-tight edge well, made by the inert-gas welding technique.

5.7   Overall interrupter design

5.7.1   *Electrical considerations*

Having described the components which go into a vacuum interrupter, the factors that must be considered in bringing them together are reviewed. These are discussed, somewhat arbitrarily, under electrical, mechanical and thermal headings. All are important to good design.

The principal dielectric stress points are the triple junctions between insulation, vacuum and metal, and also the edges of internal metal parts. Good shield design is described in Section 5.5. Increasing clearances also reduces stress, but here we encounter the constraint of economics. Generally speaking, greater clearances dictate larger interrupters. In point of fact there has been substantial progress in *reducing* the size of interrupters in recent years. This saves

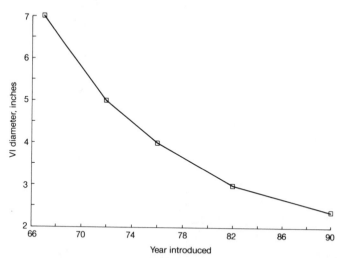

Figure 5.20   Size reduction achieved by improved design for interrupters rated 15 kV, 12 kA interrupting

(Courtesy of Westinghouse Corp.)

cost in materials, but also in processing, inasmuch as more can be handled in a given bake-out or brazing cycle. The progress made by one manufacturer in reducing interrupter size is seen in Figure 5.20. Much depends on the diameter of the contacts; surrounding components, including the envelope, can be reduced in size to the extent that the contacts can be reduced.

Nothing has been said so far about the way in which current is conveyed to the moving contact. Flexible copper braids are normally used for this function. They effectively avoid the maintenance required by sliding contacts. The inter-dependence of electrical, thermal and mechanical consideration are quite obvious in this component. Adequate current carrying capability is really a thermal matter. The braids, on the other hand, place a mechanical constraint on contact motion by increasing the effective mass to be moved and adding to its inertia. An alternative to flexible braids is thin leaves of copper; an example appears in Figure 5.21. Having an alternate path for current, through the bellows, should be avoided.

When shrinking down the size of an interrupter, one must not overlook the possibility of external flashover. There is no problem under oil or in SF_6, but most equipments are air insulated. It may be necessary, therefore, to clad the interrupter with a polymeric jacket which incorporates sheds to increase the surface flashover distance. This adds to the cost.

As an alternative, some outdoor switches have a fibreglass interrupter module housing with special coating to inhibit ultraviolet activity. The sealed vacuum interrupter chamber is encapsulated in a solid-dielectric closed cell foam to seal against moisture and other contaminants. More details of this technique is found in Section 8.4.

In addition to concern for external flashover, breaker and switch designs must guard against pole-to-pole sparkover. The simplest approach is adequate pole spacing. In these days when equipment compaction is important, this is not attractive, accordingly, some kind of interphase barrier is called for. These are sometimes cast and sometimes fabricated from sheet insulation stock. An alternative is the tubular construction shown in Figure 5.22. The tube is cut away to show the interrupter inside.

The tubular form of insulation is continued for the connections to the top and bottom of the interrupter, that is to say, tubular insulating shields protect the stabs which connect the interrupter to the primary disconnects.

A cross-section of this circuit breaker, presented in Figure 5.23, reveals another interesting feature, the use of spring loaded roller contacts to conduct the current to and from the moving shank. It is identified as item 4.

5.7.2 *Mechanical considerations*

A prime mechanical consideration is avoiding undue stress on the ceramic (or glass)-to-metal seals due to the pounding of closing and opening operations. This requires that these joints be mechanically decoupled as much as possible from the contacts and their shanks. This decoupling is assured by the bellows at the moving contact end, but must be provided by the flexibility of the end plate at the fixed contact end. This is illustrated diagrammatically in Figure 5.24.

Further mechanical support is often provided by insulating struts of epoxy-bonded glass fibre which are mounted in parallel with the interrupter. Such struts are clearly visible in Figure 5.25.

Another matter of some importance is the alignment of the contacts. Uneven erosion of the contact surface can disturb the alignment since it can cause

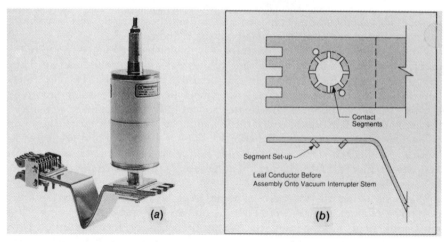

Figure 5.21   Flexible connection from moving contact to draw-out stabs

(a) *Physical arrangement*
(b) *Details of leaves*

(Courtesy of Westinghouse Corp.)

Figure 5.22   Vacuum circuit breaker with interphase insulation provided by insulating tubes

(Courtesy of Calor Emag/ABB)

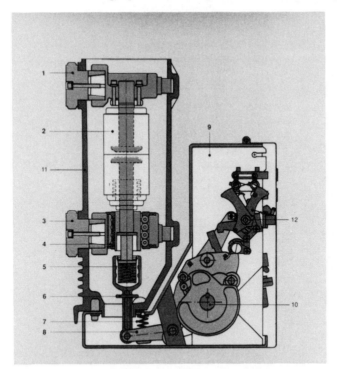

Figure 5.23   *Section of circuit breaker shown in Figure 5.21, rated for 12 kV, 2500A carrying, 31.5 kA interrupting*

(Courtesy of Calor Emag/ABB)

1 Upper connection
2 Vacuum interrupter
3 Lower connection
4 Roller contact (swivel contact for 630A)
5 Contact pressure spring
6 Insulated coupling rod
7 Opening spring

8 Shift lever
9 Mechanism housing with spring operating mechanism
10 Drive shaft
11 Pole tube
12 Release mechanism

cocking or tilting of the contacts. A guide is frequently used to minimise these effects. It is typically made of nylon and is placed over the shank of the moving contact. It is not a tight sleeve bearing, but as its name suggests, a guide to maintain the coaxility of the fixed and moving contacts.

5.7.3   Thermal considerations

Whereas the contacts of oil and gas blast circuit breakers are immersed in a fluid which can conduct and convect away from them the heat generated by ohmic and contact losses, the internal elements of a vacuum interrupter are essentially swathed in a thermal blanket provided by the surrounding vacuum. All heat must exit by the contact shanks. This fact assumes increasing importance as the steady state current rating of the interrupter is increased and as the rated voltage

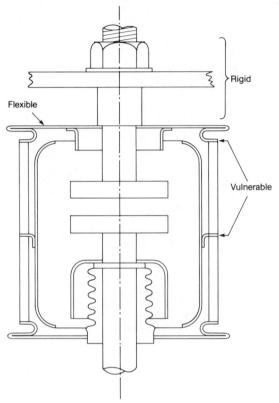

Figure 5.24   Flexibility in the end plate of a vacuum interrupter decouples the contact structure from the ceramic-to-metal joints, thereby relieving mechanical stress

is increased, since higher voltage tends to lead to physically longer interrupters and therefore longer shanks. There is nearly proportional correspondence between electrical and thermal resistance in the conduction mode. Thus, the contact interfaces constitute zones of minimal mechanical area, and hence they are regions of relatively high electrical resistance and are also impedances to the conduction of heat.

The fixed contact end is clearly more effective as a heat conduit than is the moving contact end. The one is solidly clamped to the breaker buswork (Figure 5.23), which in turn communicates with the heavy cross section, large surface area buses of the cubicle, while the moving end has the relatively high thermal impedance of the flexible braids. One design to maximise heat transfer through a flexible connection is shown in Figure 5.24. A series of flexible leaf conductors is connected to the moving interrupter stem by means of individual holes in each leaf. The hole design is a series of radial slots that, when flattened around the stem, provide eight discrete points of intimate contact for each leaf conductor (Figure 5.24(*b*)). As the stem moves up and down, the V–system flexes. It is quite common in gas blast breakers for the transfer of current from the moving contact to take place through a cluster of spring loaded sliding contacts. The design

Figure 5.25   Circuit breaker for high continuous current (4000 A) with cooling fins at the interrupter terminals

(Courtesy of Siemens AG)

shown in Figure 5.24 eliminates the wear experienced by sliding contacts and the mechanical upkeep that could be necessary for spring-loaded multi-piece designs, yet still delivers low thermal and electrical resistance because of the multi-point contact connection.

Standards have traditionally prescribed maximum temperature rise and maximum absolute temperature for the hottest spot on contacts. ANSI/IEEE C37.04-1979 [15], for example, gives 30 and 70°C, respectively, for copper. This is to assure no serious deterioration due to oxidation. Such limits have no relevance to vacuum devices, which are therefore exempt. There must, of course, be some limitation imposed by softening of braze material, but this would surely not be reached before limits were reached external to the interrupter. The same standard sets the same limits for external copper joints [16].

In 1988 the Short Circuit Testing Liaison (STL) issued a guide to the interpretation of IEC publication 56 [17], which among other things specifically addressed 'sealed-for-life' interrupters (of which vacuum is the prime example) with respect to thermal characteristic. It designates when and how heat runs should be performed. Although this is not an IEC standard, based on past experience of IEC adopting STL Guides, it may very well be incorporated in a future revision of IEC-56.

To comply with these standards it is found necessary to operate the shank or stem of the interrupter at a current density of approximately 500 A cm^2 and in the case of very high continuous current (3000 A and above) by placing substantial cooling fins that are fitted to the interrupter terminations to enhance the heat transfer to the surroundings. An example of this is seen in Figure 5.25. Breakers of this type have a continuous current rating up to 4 kA and can interrupt 63 kA at 15 kV.

The shield of a vacuum interrupter receives a considerable heat flux when the breaker is called on to interrupt a high short-circuit current. Metal vapour from the arc gives up its heat of condensation and electrons and ions convey the ionisation energy when they recombine at the shield. In an ideal interrupter the shield would be an infinite sink for such particles, but in practice as the shield temperature rises, all the impinging particles cannot condense, in scientific terms we say that the *accommodation coefficient* is less than unity. One mode of failure under these conditions occurs when the vapour density becomes so high locally that reignition is initiated. Maintaining a diffuse arc by AMF drives up the threshold for reignition, as does arc rotation in the case of a constricted arc. From a design point of view, it is important that the shield have sufficient heat capacity and thermal conductivity to absorb the short circuit heat flux without undue temperature rise. This is particularly important in applications where reclosing into a fault may occur, since the shield must survive multiple shots of energy. Its ability to rapidly get rid of this heat is limited because of the tenuous nature of its mechanical support, for the shield, like the contacts, is thermally insulated by the vacuum, unless it is an integral part of the envelope (Figure 5.1(*b*)).

5.8   References

1 LAFFERTY, J.M., BARKAN, P., LEE, T.H., and TALENTO, J.L: 'Vacuum circuit interrupter contacts', US Patent 3 246 979, 1966
2 ROBINSON, A.A.: 'Vacuum type electric circuit interrupting devices', British Patent 1 194 674, 1970
3 BARKAN, P., LAFFERTY, J.M. LEE, T.H., and TALENTO, J.L.: 'Development of contact materials for vacuum interrupters', *IEEE Trans.*, 1971, **PAS–90**, pp. 350–359
4 KURASAWA, Y., IWASITA, R., WATANABE, R., ANDOH, H., TAKASUMA, T., and WATANABE, H.: 'Low surge vacuum circuit breakers', *IEEE Trans.*, 1985 **PAS–104**, pp. 3634–3642
5 KANEKO, E., YOKOKURA, R., KOMMA, M., SATOH, Y., OKAWA, M., OKUTOMI, I., and OSHIMA, I.: 'Possibility of high current interruption of vacuum interrupter with low surge contact material: improved AG-WC', IEEE Winter Power Meeting, New York, NY, USA, 1992
6 SLADE, P.G.: 'Advances in material development for high power vacuum interrupter contacts', Proceedings of 16th international conference on Electrical contacts, PA, USA, 1992, (to be published in *IEEE Trans.* **CHMT)**
7 YANABU, S., TSUTSUMI, T., YOKOKURA, R., and KANEKO, E.: 'Recent technical development of high voltage, high power vacuum circuit breakers', Proceedings of 13th international symposium on Discharges and electrical insulation in vacuum, 1988, pp. 131–137
8 BENFATTO, T., DeLORENZI, A., MASCHIO, A., WIEGAND, W., TIMMERT, H.P., and WEYER, H.: 'Life tests on vacuum switches breaking 50 kA unidirectional current', *IEEE Trans.*, 1991, **PD–6**, pp. 824–832
9 BESTEL, E.J., GLINKOWSKI, M.T., and SALON, S.: 'Magnetic field calculation in vacuum interrupter structures with eddy current effects', Proceedings of 14th international symposium on Discharges and electrical insulation in vacuum, 1990, pp 484–488

10 BARKAN, P.: 'A study of the influence of dynamic overstressing and annealing on the fatigue life of convoluted bellows', *Isr. J. Tech.*, 1971, **9**, pp. 571–578

11 TSUTSUMI, I., KANAI, Y., OKABE, N., KANEKO, E., KAMIKAWAJI, E., and KOMMA, M.: 'Dynamic characteristics of high-speed operated, long-stroke bellows for vacuum interrupters', IEEE Winter Power Meeting, New York, NY, USA, 1992

12 RICH, J.A., FARRALL, G.A., IMAM, I., and SOFIANEK, J.C.: 'Development of high power vacuum interrupter', EPRI report EL-1895, 1981

13 GREENWOOD, A.N., SCHNEIDER, H.N., and LEE, T.H.: 'Vacuum type circuit interrupter', US Patent 2 892 912, 1959

14 SOKOLIJA, I., BURNAZOVIC, E., and HADZAGIC, V.: 'Optimal design of vacuum interrupter shield shapes', Proceedings of 14th international symposium on Discharges and electrical insulation in vacuum, 1990, pp. 84–87

15 'Standard rating structure for AC high-voltage circuit breakers rated on a symmetrical current basis', ANSI/IEEE, Table 3

16 Reference 16, Table 4

17 'Guide to the interpretation of IEC-56, 4th edn: 1987 high-voltage alternating-current circuit breakers', Short-circuit Testing Liaison, UK, 1987

Chapter 6

Design of vacuum switchgear 2: the operating mechanism and other mechanical features

6.1 Introduction

The purpose of a circuit breaker operating mechanism is straightforward enough; it is simply to close and open the contacts as required. There are, however, many factors to consider in designing such mechanisms to achieve the high performance demanded. There must be careful control of impact velocity and detailed analysis of the contact dynamics if bouncing of the contacts is to be avoided on closing. Also, the dynamics of opening must ensure minimal rebound when motion is arrested. These considerations force the need to profile the closing and opening strokes.

As a general rule, circuit breakers are required to provide *trip-free operation* which means that in the event of the breaker closing into a fault, energy is available for the breaker to open and clear the fault. This requires that closing and opening are sufficiently decoupled that one does not interfere with the other.

Circuit breakers spend almost all there useful life in the closed position, carrying current. This places some of the most stringent requirements on the contacts.

Circuit breaker mechanisms for vacuum or for any other switching technology, have requirements rarely found in mechanisms elsewhere. The loadings on latches and bearings are very high; the mechanism may not operate for weeks or months but must act instantly when called on; except for motors that may be used to charge springs or drive a compressor, bearing rotation is not continuous, but involves instead the turning of cam or the movement of a latch or pawl through a relatively small angle; undue friction can seriously affect timing, which in turn can seriously degrade performance. All these factors have substantial impacts on design.

Mechanisms for vacuum circuit breakers have additional requirements peculiar to the vacuum technology. Many of these accrue from the necessity to use butt contacts in vacuum interrupters. Oil, air and SF_6 breakers normally have multiple contact structures. These may comprise a male contact which engages by a sliding action with a number of spring loaded fingers which form the female contact. The requirement for very clean, oxide-free condition of the surfaces of vacuum interrupter contacts make such an arrangement quite unacceptable since it would lead to galling of the contacts and the formation of mechanical welds that the mechanism simply could not break. This situation makes butt contacts mandatory.

Butt contacts will still weld but, as noted in Section 6.4, appropriate selection of the contact material will assure that the welds formed can be readily broken. The most serious shortcoming of butt contact is their propensity to 'pop', forced apart by electromagnetic forces sometimes called blow-off forces, when carrying a high current in the closed position. Such popping should be avoided or at least kept to a minimum since it leads to arcing in the confined space that develops when the contacts separate. Arcing, in turn, can cause severe erosion of the contacts. When the contacts finally come together and stay together the molten surface may well form a weld which is difficult or impossible to break.

These various factors—contact bounce, rebound, welding and popping—are discussed in the ensuing sections. We show how they are factored into the design and describe operating mechanisms that satisfactorily meet the challenge that the factors impose. Because of its overriding importance and the profound influence it has on mechanism design, the subject of contact popping is addressed first.

6.2   Contact popping

6.2.1   Origin of popping forces

The actual area of contact between the butt contacts of a vacuum interrupter in the closed position is a very small fraction of the total area of the contact. Indeed, because of the macroroughened surfaces, the area is significantly less than the apparent area, since true contact only exists where the asperities of one contact bed into the surface of the other. The area will increase as the force between the contacts increases because of deformation of the surfaces. Figure 6.1 shows an idealised picture of the situation just described. On the left the current constricts through the area of contact [1]; on the right is a blow-up of this area showing a random distribution of true contact points and the envelope [2].

The magnetic field of the radial component of current converging on the constriction in one contact reacts with the radial component of current diverging from the constriction in the other contact. The result is a repulsive force tending to force the contacts apart. Put another way, the circumferential magnetic field of the axial component of the current in the contacts, reacts with the radial components of the current at the constriction to produce the repulsive force. Inasmuch as the force is proportional to the product of the current and the magnetic field, and in so far as the field is proportional to the current, the force is proportional to current squared in an ideal symmetrical situation. Holm [1] gives the following formula for the repulsive force:

$$F_r = 10^{-7}I^2 ln(b/a) \tag{6.2.1}$$

where I is the current, b is the radius of the contact, and a is the radius of the apparent contact area. The force falls to zero as a approaches b, pointing up the desirability of having as large a contact area as possible. A practical, rule-of-thumb formula for the popping force is

$$F_r = N\,100\left(\frac{I/N}{30,000}\right)^2 \text{ lb} \tag{6.2.2}$$

where N is the number of contacts. This clearly illustrates the advantage of the

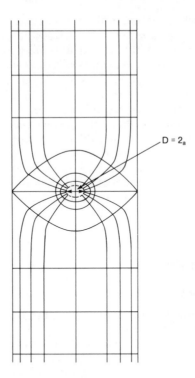

Figure 6.1(a)   Lines of current flow and equipotential surfaces of current constriction in ideal contact [1]

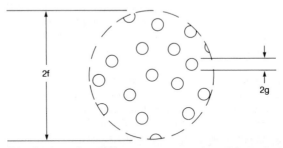

Figure 6.1(b)   Distribution of multiple contact spots and envelope enclosing the spots [2]

multiple finger contacts enjoyed by other switching technologies. For vacuum, with butt contacts, N is usually one, thus a current of 30 kA produces a popping force of 100 lb (225 Newton).

The force depends on the instantaneous current. If a fault current of 50 kA RMS occurs, it can have an asymmetrical peak, if fully offset, of $2\sqrt{2}$ times this value, in which case the repulsive force according to Equation 6.2.2 is

$$\left[\frac{2\sqrt{2}\,50}{30}\right]^2 \times 100 = 2,220\,\text{lb}$$

and this is for one pole only. As mentioned already, the mechanism must provide a counter force if popping and its deleterious effects are to be avoided.

Rich *et al.* [3] point out the crucial effect of contact force on the design of a spring-actuated mechanism. It substantially influences the following:

(i)   latch force and the number of toggle linkages required in series for tripping, which, in turn determines the effective mass of the mechanism
(ii) bearing size and the frictional energy loss in bearings
(iii) size of the closing springs
(iv) number of stages in the gear box and size of the motor used to charge the springs.

In addition, excessive energy in wipe springs* necessitates use of an opening dashpot to control the impact and the rebound of the contacts and excessive energy in closing springs also results in a large amount of leftover kinetic energy in the system. This requires a closing dashpot or brake for its absorption. What we are witnessing here is an effective chain reaction on the size, energy, and ultimately, on the cost of the mechanism. It is therefore essential that the nature of these forces be thoroughly understood, and the best possible trade-offs between simplicity, reliability, cost and size be selected in mechanism design.

The formula of Equation 6.2.2 is idealised. In practice there will be a number of factors that influence the popping force. These include the large variation from operation to operation in the distribution of contact points on the rough contact surfaces. One would expect a statistical variation of popping force as a function of current following repeated closing operations. Rich *et al.* [3] set up the apparatus shown in Figure 6.2 to investigate this variation. Current could be

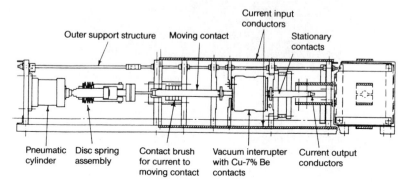

Figure 6.2   Experimental set-up to measure contact popping forces in vacuum interrupter [3]

(Courtesy of EPRI)

passed through the contacts of a vacuum interrupter and a holding force could be applied between the contacts. The mechanical force exerted on the closed contacts was developed by a standard pneumatic operating cylinder. The

---

* The wipe springs are those which are compressed following contact mating. Their gradient and the contact travel to the fully closed position after mating, determine the counter popping force.

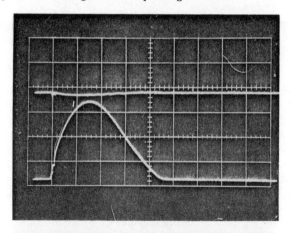

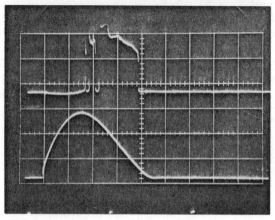

Figure 6.3   Oscillograms showing voltage (upper trace) and current (lower trace) taken during high momentary test [3]

(a) *No contact popping*
Scale: 1 div. = 10 V = 20 kV = 2ms
(b) Contact popping
Scale: 1 div. = 10 V = 40 kA = 2 ms

(Courtesy of EPRI)

contacts were essentially isolated from the mass and stiffness characteristics of the operator by means of a special coupling employing stacked disc springs which closely simulated the character and function of the wipe spring conventionally used with vacuum interrupters. The static mechanical force exerted on the contacts was determined from the deflection of the disc spring stack which had been previously calibrated on an Instron machine. Precautions were taken to assure that friction and contact binding did not affect the results in any significant way.

A test series was started by applying considerable force which was then reduced in increments while the current for each test was kept constant until a point was reached where contact popping was observed. The incidence of popping was evident from a measurement of the voltage across the contacts. Popping created an arc whose presence was immediately manifest by the appearance of an arc voltage. Tests with and without popping are shown in Figures 6.3(*a*) and (*b*). The current in this instance was supplied from a charged capacitor bank. Sufficient testing was done to assure validity of the subsequent statistical analysis.

6.2.2   Magnitude of the popping force and implications for mechanism design

From what has been said about the origin of the popping force, the force *F* necessary to suppress popping should be proportional to the square of the current

$$I^2/F = C \qquad (6.2.3)$$

where *C* is a constant. As expected, results from tests using the equipment of Figure 6.2 [3] show statistical variations (Figure 6.4). The variation in *C* in the test data is a reflection of the presence of variables which are not controlled or are not controllable and which produce an uncertainty in the result. Figure 6.4 can be interpreted as the level of contact force required for a given current to prevent popping with various degrees of reliability. The data were obtained from

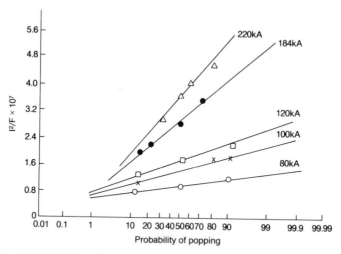

Figure 6.4   Statistical plot of contact popping test data [3]
   (*Courtesy of EPRI*)

an interrupter with Cu–7% Be contacts. This is a material which has been shown to have exceedingly good switching performance [4]. To the best of my knowledge, it has never been used commercially because of its toxicity. More conventional contact materials will give similar data.

It is important to define the degree of reliability needed from an engineering design viewpoint. Frequent popping is to be avoided for the reasons stated and excessive contact forces are also undesirable. The maximum current that interrupter contacts experience occurs when a fully asymmetrical fault current is generated at the full breaker rating. Both of these are rare events individually and statistically. The simultaneous combination of these events is even more unlikely. If these two events are further combined with a design which produces only a small but finite chance of popping under high-current conditions, the result will be that popping can be eliminated for all practical purposes. Because of the statistical nature of the popping phenomenon, an engineering trade off has to be made between the force and energy requirements of the mechanism and the degree of reliability required to reduce the risk of damaging the dielectric strength of the interrupters. Past experience and the data of Figure 6.4 suggest that it is reasonable to accept a force level which results in 95% reliability.

Entering the curves of Figure 6.4 at the 95% reliability level, it is possible to

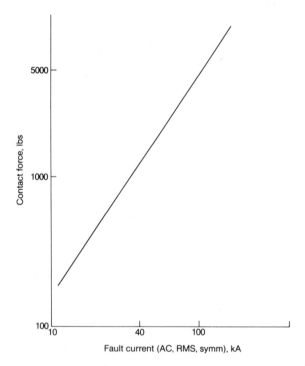

Figure 6.5   Contact-popping force as a function of fault current (experimental results, force
$=C_1^{1.45})$ *[3]*

(Courtesy of EPRI)

develop a plot of contact force versus fault current; this has been done in Figure 6.5. The results show that the required force does not follow a simple square law relationship with current, indicated in Equation 6.2.1, but that it can be represented by the expression

$$F = C_1 I^{1.45} \tag{6.2.4}$$

where I is the peak fault current. For the holding force in pounds, $C_1 = 5.3 \times 10^{-5}$. Reducing the reliability to 75% and accepting the correspondingly higher incidence of popping would require substantially less contact holding force.

To summarise, electromagnetic forces tend to blow apart butt contacts when the contacts carry a high current. A constraining force must be applied to prevent contact separation, since such popping causes arcing which can seriously degrade the interrupter. The constraining force is important as it materially affects many aspects of the design. It can be kept within reasonable bounds if popping is accepted on very rare occasions.

The electromagnetic forces considered so far derive from currents within the contact structure itself. Consideration must also be given to the electromagnetic effects of buswork which brings the current to and from the interrupter. The forces generated in this way are usually small compared with those already discussed.

6.2.3   Providing the contact holding force

The usual way of providing contact holding force is to place a spring in the driving linkage, relatively close to the interrupter. The wipe spring (as it is called), is precompressed so that immediately after the contacts meet a significant force is developed between them which increases as the mechanism moves to the final latched closed position. The mechanism travels further before contact touch. As the contacts wear and erode during their normal duty, the compression of the spring is less and the contact force is diminished. The precompression helps minimise this reduction.

Figure 6.6 shows that location of the wipe springs (identified as item 10) in one circuit breaker design. They are placed adjacent to the bell-crank mechanism immediately below the interrupters. In some contactors and switches the wipe springs are put within the movable contact which is hollow. In this way they exert their force directly on the contacts. They are not within the vacuum envelope, of course, but are inserted during assembly of the contactor. The rather busy drawing taken from US patent 4,568,804 shows an example; the wipe spring is item 264 (Figure 6.7).

A *magnetic assist* is sometimes applied to contactors. This is a magnetic means whereby the current flowing through the contacts is used to oppose the repulsive popping force. Such devices are quite simple. They usually comprise a single turn of the current cable which links a magnetic circuit. The magnetic field produced pulls on an armature which applies closing pressure to the contacts through a lever. The greater the current, the greater will be the force.

Contactors regularly see overcurrents during motor starting. Inrush currents are typically six times normal and can have appreciable asymmetry on at least two phases. Since this is a frequent duty, the contactor must not experience contact popping when inrush currents occur. More serious overloads such as faults are a different proposition; the consideration here is safety. When a short circuit occurs, no interrupter of the contactor must explosively self destruct. The

Figure 6.6   Underside view of vacuum circuit breaker showing wipe springs for contact loading

(*Courtesy of Westinghouse*)

1 Breaker code plates	8 Ground disconnect
2 MOC operator	9 Operating rod
3 Floor tripper—trip	10 Contact loading spring
4 Levering latch	11 Spring yoke
5 Floor tripper—spring release	12 Bell crank
6 Secondary disconnects	13 Primary disconnect
7 TOC operator	14 Pole base

contactor must not, for example, blow its cover. However, standards [5] do not require that the device be serviceable after such an event. This is not a particularly difficult requirement for a contactor protected by a fuse because the fault duration is short. Where a breaker is used for protection, the contactor must be able to sustain eight cycles of arcing [5]. This is not trivial. With an arc voltage of 30 V and a current of 10 kA RMS, the power is 300 kW and the energy released is 40 kJ! The chief recipient of this energy is the interrupter shield which ablates under the intense heat flux. One contactor manufacturer solves the problem by incorporating sufficient mass in its stainless steel shield to accept the abuse of the arcing.

By their nature, axial magnetic field contact structures produce a magnetic assist; the magnetic field of the current in one contact produces a south pole which looks at the north pole produced by the current in the other contact. The attractive force tends to increase with the square of the current. It is possible to

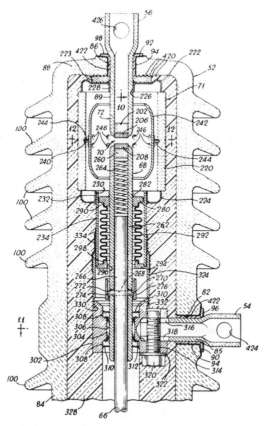

Figure 6.7   *Vacuum interrupter with wipe spring located within contact*

   (*Courtesy of Joslyn Hi-Voltage Corp.*)

compute the magnitude of this force by D'Alambert's principle, sometimes referred to as the principle of virtual work. The procedure is to equate the work done by moving the contacts a short distance Δx against the force, to the increase in stored energy

$$F\Delta x = \left(\frac{B^2}{2\mu_0}\right)(A\Delta x)$$

or

$$F = \frac{B^2}{2\mu_0}\,A \tag{6.2.5}$$

If the area A is in square metres, and the flux density B, is in webers per square metre, the force will be in newtons. μ_0 is the permeability of free space $(4\pi \times 10^{-7})$. Some uncertainty will exist regarding the area to use over which B is assumed constant.

6.3 Contact bounce and rebound

6.3.1 Contact dynamics

An appropriate introduction to this topic is to quote from Barkan [6]: 'When two finite bodies with initially different velocities collide, it is impossible for their interface to remain in contact, unless some measure of dissipation of the initial kinetic energy occurs. This is so, because it is not possible to find a common velocity for the interface which simultaneously satisfies the laws of conservation of energy and momentum. Thus, to maintain contact between two surfaces it is necessary that a certain critical portion of the incident kinetic energy be dissipated'. This clearly has relevance to the situation which occurs when the moving contact strikes the fixed contact during the closing of a vacuum switching device. If the contacts do not remain in contact, they bounce.

Contact bounce is undesirable because, like contact popping, it allows arcing when the contacts separate, with all the problems described in Section 6.1.

The fundamentals of impact between two bodies, the simple system with one degree of freedom, will be found described in any basic text on dynamics. The contact structure of a switch is considerably more complicated but much can be learned from the simple case. Shames [7] considers the period of collision to be made up of two subintervals of time. The *period of deformation* refers to the duration of the collision starting from the first contact of the bodies and ending with the time of maximum deformation. The second period, covering the time from the maximum deformation condition to the condition in which the bodies just separate, he refers to as the *period of restitution*. He goes on to define the *coefficient of restitution* as the ratio of the impulse during the restitution period and the impulse during the deformation period, i.e.

$$\varepsilon = \frac{\text{impulse during restitution}}{\text{impulse during deformation}} = \left(\int R dt \middle/ \int D dt \right)$$

where R and D are, respectively, the impulsive forces between the bodies during restitution and deformation.

By equating the impulse experienced by both bodies to the change in momentum of each body in both the period of deformation and the period of restitution, and noting that during the period of contact the bodies have the same velocity, Shames [7] derives an alternative expression

$$\varepsilon = - \frac{\text{velocity of separation}}{\text{velocity of approach}}$$

The coefficient ε depends markedly on the material properties of the colliding bodies and on the size, shape and velocities of the bodies before impact. In a collision between two *perfectly elastic bodies* ($\varepsilon = 1$), the velocity of separation is equal to the velocity of approach. If one of the bodies is so massive as to be considered immovable, the second rebounds with the same velocity as it strikes (the ball bearing bouncing off the heavy steel plate). In both instances no energy is lost. At the other end of the scale, where one or both bodies is made of a putty-like material, ε approaches zero. In this extreme the bodies move in unison after the impact. In the second example with the very massive body, they do not move at all.

In more realistic situations one must expect something between these extremes. During the period of deformation kinetic energy is converted into strain energy within the bodies. Because the impact is inelastic, some of this energy is not reconverted to kinetic energy during the period of restitution but is dissipated instead as heat (and noise) in the bodies and their environment and the bodies are permanently distorted. As noted earlier, if a certain critical amount of the initial kinetic energy is dissipated, bounce can be avoided.

6.3.2   Controlling contact bounce

All methods of contact bounce control are concerned with minimising the portion of the incident kinetic energy that must be dissipated and maximising the rate of dissipation. The fixed contact of a vacuum interrupter is relatively rigid so much of the deceleration of the moving contact is accommodated by distortion of the impacting surfaces rather than by acceleration of the 'fixed' contact. Thus, following what has been said above, some of the incident kinetic energy will be dissipated directly at the interface in the immediate vicinity of the region of contact because of the inelastic behaviour of the contact material. However, a real contact assembly is not a simple, single degree of freedom system, ideally isolated from the support structure. More generally, contacts are directly coupled to a supporting structure which may profoundly influence their post-impact behaviour.

As Barkan [6] points out, it is instructive to consider a relatively simple case which demonstrates the influence of the stationary support, on the separation of the impacting members. Even when the coefficient of restitution is zero, it is possible to encounter contact separation. Assume a fixed contact to be

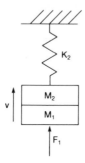

Figure 6.8   Model for deriving critical stiffness of contact mounting [6]

   (By permission of IEEE)

represented by stiffness K_2 and mass M_2 in Figure 6.8. Impact occurs as M_1 with initial velocity V_{10} strikes stationary mass M_2 which is initially motionless. Assuming all of the strain energy produced by the impact is completely dissipated $(\varepsilon = 0)$, the two masses $M_1 + M_2$ will move together at velocity $v = v_{10}(1 + M_2/M_1)^{-1}$ which is obtained from the conservation of momentum. The combined mass, $M_1 + M_2$, together with the support structure spring gradient K form an oscillating system which will oscillate with an amplitude X

which is easily found to be

$$X = \frac{M_1}{M_1 + M_2} \frac{v_{10}}{\omega} \sin\omega t + \frac{F}{K}(1 - \cos\omega t) \tag{6.3.1}$$

where $\omega = [K(M_1 + M_2)]^{1/2}$ and F is the driving force of M_1. M_1 will separate from M_2 if at any time during the oscillation the following condition is encountered

$$\ddot{X} < -F/M_1 \tag{6.3.2}$$

By differentiating Equation 6.3.1 twice and substituting in Equation 6.3.2, the following restriction on the stiffness K of the contact support structure is obtained

$$K < [(1 + M_2/M_1)^2 - 1][1 + M_2/M_1]F^2/M_1 v_{10}^2 \tag{6.3.3}$$

Equation 6.3.3 then defines a critical stiffness for the mounting of the

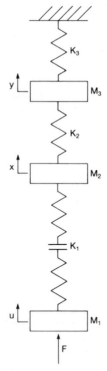

Figure 6.9   Three degrees of freedom system for energy transmission models [6]
        (*By permission of IEEE*)

stationary contact which, if exceeded, may produce contact separation.

Although Equation 6.3.3 is a necessary condition for the suppression of bounce it is not a sufficient one because the energy dissipating ability of most contact materials is not 100% as has been assumed in the analysis. Since the energy dissipation at the contact interface will not be adequate to prevent contact bounce, the principal recourse available is to exploit the support structure as an energy sink.

Barkan [6] uses the model shown in Figure 6.9 to analyse the transfer of energy to the support structure. In this figure, contact M_1 initially moving with velocity v_{10} and driven by force F impacts the stationary contact M_2, mounted on a support structure represented by mass M_3, and initially unloaded springs K_2 and K_3. K_1 represents the equivalent linearised stiffness of the interface between contacts M_1 and M_2. The principal conclusions which can be drawn from these results is that the optimum situation is obtained when $K_2/K_1 = 1 = M_3/M_1$. Under these conditions, a maximum percentage of the incident kinetic energy is transmitted into and maintained within the support structure during the critical impact period.

A certain amount of energy will inevitably go into frictional rubbing and shearing at the surface. The remainder of the transferred energy which passes to the fixed contact will promulgate through the support structure as a travelling wave. Bouncing is minimised by endowing this transmission line with a high damping characteristic and having it dissipate energy while the contacts remain closed. This requires that the dominant natural mechanical frequency of the fixed structure be much higher than that of the moving contact. Schemes for rapid dissipation of the transferred energy involve the use of dissipative media with high internal losses, rubbing friction, and secondary impacts with auxiliary components. However, it is generally found in vacuum switching equipments that special steps are not necessary, that the system has sufficient inherent losses.

Some contact materials, copper/bismuth is an example, are relatively soft in the annealed state wherein they find themselves following interrupter bakeout. This condition favours a low coefficient of restitution which is good from an antibounce point of view. Mechanical pounding of the contacts by repeated closing operations tends to work harden the surfaces where they impact, thereby increasing the value of ε. This may cause bouncing where it did not exist before. Unnecessary mechanical operations should therefore be avoided. Opening operations under load cause modest arcing which tends to ameliorate the situation by softening the contact surfaces to some extent.

A problem similar to bouncing is that of rebound on opening. The moving parts of the circuit breaker, contactors or switch must be stopped at the end of the opening stroke. It is important that much of the kinetic energy be absorbed in the impact to avoid a rebound that would bring the contacts close together again. The dielectric recovery of the switch is put at risk if the rebound is too great.

Building some dissipative element into the arresting buffer is an adequate solution in many instances. Some circuit breakers employ an air or oil dashpot. A typical dashpot comprises a hydraulic cylinder containing silicon oil; the cylinder is made of steel and the piston is plastic. The oil is expelled through an orifice, producing a damping force approximately proportional to the speed. The dashpot should be located close to the centre of gravity of the moving masses and this typically means that the piston is driven from a crank on the common bar which rotates to drive the three poles during the opening operation. Barkan [6] has some helpful comments on the use of dashpots to control deceleration.

6.4   Contact welding

The inevitability of contact welding was discussed in Section 6.1. It was pointed out that the consequences of welding are beneficial from the viewpoint of current carrying, but that one must be able to break the weld readily when the contacts are called on to open.

The almost universal way of breaking contact welds is by impact. When the switch closes, the mechanism continues to travel after the contacts meet to compress the wipe springs and provide the necessary force between the contacts to avoid popping under high momentary current conditions (Section 6.2). It follows that when the switch opens, the mechanism travels the same distance before the contacts part. By the instant of would-be contact parting a substantial amount of kinetic energy is already stored in the moving parts of the mechanism. This energy is used to fracture the weld by snatching the moving contact, which up to this time has been mating with and welded to the fixed contact, and by pulling it away from its mate. The action is impulsive.

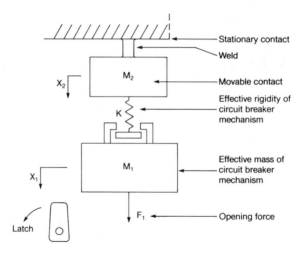

Figure 6.10   Circuit breaker model for weld breaking analysis [8]

(By permission of IEEE)

Figure 6.10 shows a simple analytical model for the system which is applicable to the sequence just described. This figure and the following analysis derive from Reference 8.

Nomenclature

F_1 = *force of opening spring*
M_1 = *mass of mechanism*
M_2 = *mass of movable contact*
K = *equivalent rigidity of circuit breaker mechanism*
F_w = *weld breaking force*

v = velocity
x = displacement
δ = total stretch associated with strain in weld
δ_f = weld stretch to fracture

This analysis considers the influence of weld characteristics upon the initial contact motion during separation in a circuit breaker. Figure 6.10 illustrates a model employing all of the key elements. Assuming the idealised weld characteristics of Figure 6.11, either ideally plastic or ideally brittle–elastic

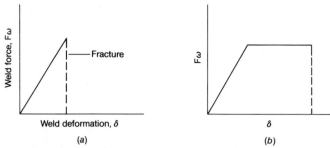

Figure 6.11   *Force-deformation characteristic [8]*

    (a) *Ideally brittle welds*
    (b) *Ideally plastic welds*

    (*By permission of IEEE*)

weld characteristics may be readily considered. In the ideally brittle–elastic case no yielding is present. Failure of the weld occurs instantaneously when the weld strength is exceeded. In the more critical, ideally plastic case, the weld yields at constant force until failure occurs at $\delta = \delta_f$.

A useful simplification assumes that the weld stiffness is much greater than K. The equation of motion can then be divided into three principal intervals.

(i)   The mass M, initially at velocity v_{1i}, impacts M_2, stressing the linkage up to a force level F_w corresponding to the contact weld yield strength. At the end of this interval, the velocities of the two elements M_1 and M_2 are approximately

$$v_{10}=\sqrt{v_{1i}^2-F_w^2/KM_1} \tag{6.4.1}$$

$$v_{20} = 0 \tag{6.4.2}$$

(ii) (a) For the ideally brittle–elastic case Equations 6.4.1 and 6.4.2 define approximate conditions at the instant of weld fracture. It can be shown that subsequent contact motion comprises a series of decaying oscillations about the mean motion defined by conservation of momentum. Thus the mean contact velocity following-impact is

$$v_{12} = \frac{M_1}{M_1 + M_2}\, v_{10} \tag{6.4.3}$$

(b) For the ideally plastic case, weld fracture does not occur until $x_2 \geqslant \delta_f$.

During this second interval the equations of motion are

$$\ddot{X}_1 + (K/M_1)(x_1 - x_2) = F_1/M_1$$
$$\ddot{X}_2 - (K/M_2)(x_1 - x_2) = -F_w/M_2 \tag{6.4.4}$$

as in Equations 6.4.1 and 6.4.2 at $\ell = o$:

$$\dot{x}_1 = v_{10}, \ \dot{x}_2 = x_2 = 0, \ x_1 = F_w/K$$

For simplification in an analysis erring on the side of caution, assume $F_1/M_1 \ll F_w/M_2$. With this further restriction, the solution to Equation 6.4.4 is

$$x_2 = (Kv_{10}/M_2\eta^3)(\eta t - \sin\eta t) - (F_w/2(M_1 + M_2)\eta^2)(\eta t)^2 \tag{6.4.5}$$

$$\dot{x}_2 = (Kv_{10}/M_2\eta^2)(1 - \cos\eta t) - F_w\eta t/(M_1 + M_2) \tag{6.4.6}$$

$$\dot{x}_1 = v_{10}/(1 + M_2/M_1)\cos + (F_1 - F_w)/(M_1 + M_2)t$$
$$+ v_{10}M_1/M_1 + M_2 \tag{6.4.7}$$
$$\eta^2 = K(M_1 + M_2)/M_1M_2$$

Of particular interest are the velocities of the two bodies at the instant of weld fracture which occurs when $x_2 = \delta_f$. From Equation 6.4.5, ηt_f is found. Then, $\dot{x}_{2f}$, $\dot{x}_{1f}$ in Equations 6.4.6 and 6.4.7 may be evaluated. As a criterion of the adequacy of the behavior, we require that for satisfactory response,

$$(M_1\dot{x}_{1f} + M_2\dot{x}_{2f})/(M_1 + M_2) > v$$

Here v represents the minimum desired velocity for contact separation.

(iii) Once the weld is broken the subsequent contact motion can be calculated by the usual dynamical relationships.

6.5   Spring-activated, cam-follower mechanism

6.5.1   Basics

The purpose of a breaker operating mechanism is to close and open the breaker on command in accordance with desired closing and opening speed profiles. Figure 6.12 shows examples of such profiles. A charged spring is a simple and reliable way of storing the energy for such operations and is well adapted to applications requiring the delivery of relatively high power for short intervals, which coincides with the needs of a vacuum breaker. However, if the movable contact is directly coupled to the spring, the motions shown in Figure 6.12 are not readily achieved. This difficulty can be resolved by interposing a cam between the spring operator and the follower mechanism.

A typical mechanism of this class is presented schematically in Figure 6.13,

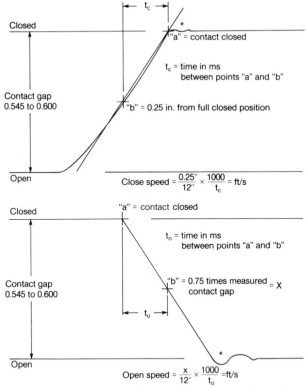

Figure 6.12 Closing and opening operating profiles for spring-activated cam-follower mechanism

* For breakers equipped with an opening dashpot, there is virtually no overtravel or rebound

(Courtesy of General Electric Co.)

which shows a follower system of mass M_c and stiffness K which is to be moved against a load F_y by a spring-activated system. This, in turn, is represented by a cam-flywheel of inertia I_α which is driven by a spring. The spring acts through a crank connected to the cam. The cam is initially constrained by a latch in a position α_0 past the bottom dead-centre position of the crank arm to which the spring is attached. The system is operated by releasing the latch and allowing the spring-crank drive to rotate the cam, displacing the follower to an upper latched position. The excess kinetic energy in the cam causes it to rotate slightly past the top dead-centre position, thereby partially compressing the spring. A suitable drive then slowly completes the revolution, precharging the spring for the next revolution.

The critical feature is the cam profile. Barkan and McGarrity [9] point out that because the cam rotational velocity is neither constant nor known and because of the strong interaction between the dynamics of the cam and follower in such systems, conventional cam design techniques are not applicable. Nevertheless, they show that long-hand calculations and a simple graphical

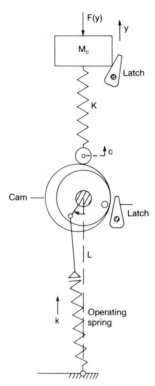

Figure 6.13 Schematic diagram showing principal members of simple spring-actuated, cam-follower mechanism [9]

technique suffice to design a system to meet prescribed dynamic performance characteristics and they go on to make an experimental check on a specific design, which shows satisfactorily agreement.

6.5.2 *Spring-activated mechanism for three-pole circuit breaker*

Almost without exception the principal manufacturers, worldwide, of vacuum switchgear have elected to adapt the basic idea outlined in the last section as the basis for their breaker mechanism design. The products of various manufacturers differ in detail but not in overall concept. The ML-18 mechanism of General Electric is described by way of example. That company's permission to allow this is gratefully acknowledged.

The ML-18 mechanism (Figure 6.14) uses a gear motor to charge its closing spring. During a closing operation, the energy stored in the closing spring, which is typically in the range of 150–200 joules, is used to close the vacuum interrupter contacts, charge the wipe springs which load the contacts (Section 6.2.1), charge the opening springs, and overcome bearing and fictional forces. The energy then stored in the wipe and opening springs will open the contacts during an opening operation.

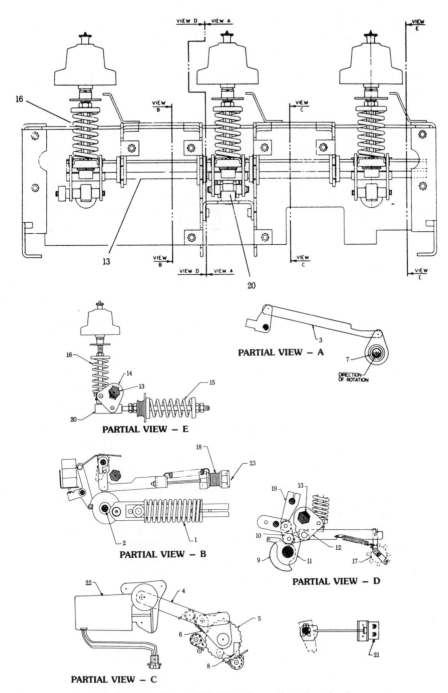

Figure 6.14   Schematic diagram of ML-18 spring activated mechanism
   (Courtesy of General Electric Co.)

Closing and opening operations are controlled electrically by the control switch on the metal-clad door or by remote relaying. Mechanical control is provided by manual close and trip buttons on the circuit breaker. The closing spring may be manually charged and a method for slow closing the primary contacts is available when the circuit breaker is withdrawn from the metal-clad cubicle.

6.5.2.1   Closing spring charging

Figure 6.14 shows a front view of the ML-18 in a schematic form. The primary contacts are open and the closing spring (1, view B) can be seen mounted on the left side of the breaker and the electrical charging system mounted on the right side of the breaker. Both components are fastened to the cam shaft (2, view B). A manual charging system (3, view A) is provided so that the mechanism can be slow closed and the closing spring can be charged if there is a loss of electrical control power.

Spring charging is accomplished electrically by a rotating eccentric on the output shaft of the gear motor driving pivoted charging arms (4, view C) which oscillate about the centreline of a ratchet wheel (5, view C). A driving pawl (6, view C), mounted within the charging arms, oscillates with the charging arms. Starting from its rear–most position, as the charging arms rotate forward, a spring forces engagement of the driving pawl with a tooth on the ratchet wheel. The ratchet wheel is advanced by the rotating charging arms and pawl assembly. Advancement of one tooth spacing is provided for each oscillation of the system. The ratchet motion is restricted to one direction by a spring-loaded holding pawl that prevents the ratchet wheel from going backwards as the charging arms oscillate back to pick up the next tooth. Thirteen complete cycles of the charging arms are needed for a full charge of the closing spring. The efficient, compact gear motor accomplishes this action in about two seconds. When the charging cycle is complete, the ratchet wheel is positioned so that a missing tooth is adjacent to the driving pawl and any motor overspin will not drive the ratchet wheel, thus preventing damage to the system.

When the spring is completely charged, the assembly is retained in that position by the close latch, until it is desired to close the circuit breaker.

The manual charging system (3, view A) works directly on the cam shaft where a one-way clutch (7, view A), driven by a manual handle, provides rotation of the ratchet wheel. Manual pumping of the handle advances the ratchet wheel and the holding pawl prevents counter-rotation while the handle is returning for another stroke. Approximately eight complete strokes of the manual handle are required for one complete spring-charging operation.

6.5.2.2   Closing operation

By either energising the close solenoid or depressing the manual close button, the close latch (8, view C) is rotated, releasing the closing spring (1, view B). This action releases the energy in the closing cam (9, view D) and closing roller (10, view D) and causes the linkage to rise until the prop (11, view D) can slip under the close roller (10, view D) and hold the linkage in place. As the linkage moves, the output crank (12, view D) rotates the cross shaft (13, view D) which in turn rotates the phase bell cranks (14, view E) on all three poles. The rotation of the phase bell cranks compresses the two opening springs (15, view E) on poles 1 and 3, closes the vacuum interrupters, and compresses the wipe springs (16, view E) on each pole. The rotation of the cross shaft (13, view D) also changes the auxiliary switch (7, view D) position. The position flag on the front panel will then indicate 'closed'. After the breaker is closed, the charging motor is again energised and the closing spring is charged as described under 'closing spring charging'. Spring charging is possible when the breaker is in the closed position because the linkage is held in place by the prop.

6.5.2.3   Opening operation

By either energising the trip solenoid (18, view B) or depressing the manual trip button (23, view B), the trip latch (19, view D) is rotated, permitting the linkage to collapse and the vacuum interrupter contacts to open under the force of the wipe springs (16, view E) and opening springs (15, view E). At the end of the opening stroke, the centre phase wipe spring assembly hits a stop on the frame that limits over-travel and rebound. Rotation of the cross shaft from the closed to the open position operates the auxiliary switch (17, view D) which opens the trip coil circuit. If the closing spring has been recharged, the linkage will be reset and the trip latch will be in place on the trip roller, ready for another closing operation. If the closing spring has not been recharged, the trip latch may be held out of position. A latch-checking switch (21, view C) will not close unless the latch is in its normal position. The contacts of this latch-checking switch are in the closing circuit so that electrical closing is blocked when the trip latch is not reset.

6.5.2.4   Trip-free operation

The linkage is mechanically trip-free in any location on the closing stroke. Electrically energising the trip coil while closing will, after the auxiliary switch contacts position, rotate the trip latch and permit the circuit breaker to open fully. The linkage will reset in a normal open operation and the closing spring will recharge as described in Section 6.5.2.1. Figure 6.15 is a photograph of the mechanism just described with the principal parts identified. It is actually a view from underneath, looking up (note the caster, top centre).

It only remains to connect the interrupters to the mechanism; Figure 6.16 shows an example of how this is done. The connecting rod (part 4) in Figure 6.16 is an epoxy casting, it is a variation on those shown in Figure 6.14. It is necessary, of course, to use an insulating material since the interrupter is at line potential whereas the mechanism is grounded. The bell shape is re-entrant on the underside so as to give a long and protected path against surface flashover. The operation of the wipe spring to exert pressure between the contacts when the crank (item 6) turns and the contacts are closed, is evident. As contacts erode and wear, the wipe spring will take up the slack with little change of contact force because of its precompression. The extent of the loss of contact material will be apparent from the position of the erosion indicator disc (item 15), with respect to a fixed reference. This is seen clearly in Figure 6.14.

There has been a considerable evolution in the design of spring-actuated, cam-

Figure 6.15   The ML-18 mechanism view from underneath

(*Courtesy of General Electric Co.*)

1 *Closing spring*	10 *Pivot bolt*
2 *Opening springs*	11 *Interlock bracket*
3 *Auxiliary switch*	12 *SM/LS motor control switch*
4 *Spring charging motor*	13 *LCS latch checking switch*
5 *Trip coil*	14 *CL/MS close latch monitor switch*
6 *Close coil*	15 *Stationary aux. switch operator*
7 *Ratchet wheel*	16 *Close latch adjustment screw*
8 *Closing cam*	17 *Close linkage pilot*
9 *52Y relay*	

follower mechanisms in recent years with a strong trend to simplification by reducing the number of parts and the number of points of adjustment. A typical breaker, the entire unit with three poles and the interrupters, now has about 150 parts. This compares, for example with about 400 parts for the minimum oil breaker it may have replaced. Similarly, there are typically three points of adjustment compared with 20 or more for some other technologies.

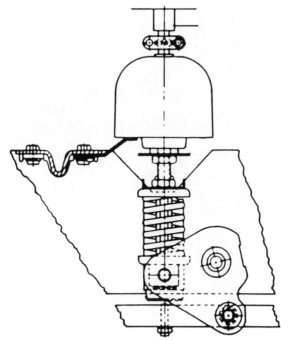

Figure 6.16   Examples of connecting rod
(Courtesy of General Electric Co.)

6.6 Four-bar linkage

Although cam-follower mechanisms are used for most vacuum circuit breakers, they are not universally selected for all vacuum switching devices. The four-bar linkage has been, and remains, a very popular mechanism for high voltage gas and oil breakers [10]. Its application to a vacuum recloser is now described.

A recloser is a self-controlled device that protects distribution circuits and equipment. Fault current sensing is provided by the control which actuates the recloser. Reclosers trip on overcurrent and then reclose automatically. If the overcurrent is temporary, the automatic reclose restores normal service. If the fault is permanent, a preset number of trip and reclose operations are performed to lockout. All three phases open, reclose and lockout simultaneously. Opening sequences can be all fast, all delayed, or any combination of fast operations followed by delayed operations, usually up to a total of four. Fast operations clear temporary faults before branch-line fuses can be damaged. Delayed operation allow time for fuses or other downstream protective devices to clear so that permanent faults can be confined to smaller sections of lines. From this description one can expect that reclosers will operate quite frequently in circuits subject to lightning and/or wind storms.

An external view of a typical vacuum recloser is shown in Figure 6.17, while Figure 6.18 shows a more detailed view when the tank has been removed. The tank normally contains oil which provides insulation across the interrupters

Figure 6.17   Vacuum recloser

(Courtesy of Cooper Power Systems)

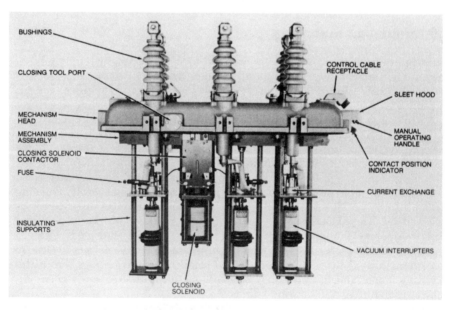

Figure 6.18   Untanked view of vacuum recloser

(Courtesy of Cooper Power Systems)

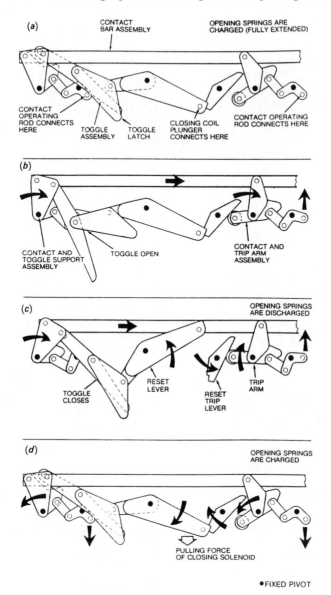

Figure 6.19   Linkage mechanism of vacuum recloser

    (a) *Contacts closed*
    (b) *Opening springs released*
    (c) *Contacts fully open*
    (d) *Contacts closed*

    (*Courtesy of Cooper Power Systems*)

when their contacts are open and isolation from ground at all times when the
bushings are energised. It plays no direct part in current interruption.

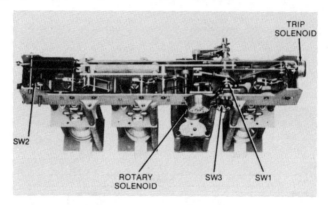

Figure 6.20   View of the vacuum recloser mechanism from top

(Courtesy of Cooper Power Systems)

The mechanism about to be described is mounted within the 'head' of the recloser. It is illustrated diagrammatically in Figure 6.19, or more correctly, Figure 6.19 shows sufficient of the mechanism to drive two poles of the recloser. The linkage continues to the right, connecting with the third pole. The black solid circles in this diagram represent fixed pivot points.

Contact opening is initiated by an electrical signal to a trip solenoid which acts on the toggle latch to release charged opening springs. These springs are not shown in Figure 6.19, but are clearly visible on the left in Figure 6.20 which shows a photograph from the top with the head removed. They are fully extended when the contacts are in the closed position, Figure 6.19(a). When the push rod of the solenoid acts on the toggle latch to open the toggle, the contact arm assemblies are rotated clockwise on their fixed pivots, causing the contact bar assembly to move to the right, Figure 6.19(b). Simultaneously, the cranks to which the contact rods are connected rotate counter–clockwise, pulling the rods upward and opening the contacts (note the upward arrow to the right in Figure 6.19(b) and (c). In the same motion the trip lever is rotated to snap the toggle closed. This motion of the reset lever also pulls the plunger out of the closing coil. At this point the mechanism is in the open (tripped) position, Figure 6.19(c)).

A signal to the rotary solenoid closes the contactor, thereby energising the high-voltage coil of the closing solenoid. As the plunger is drawn down into the coil, the reset lever is pulled down and latched, the interrupter contacts are closed, and the closing springs are charged, Figure 6.19(d). The mechanism is then ready for another opening operation.

6.7   Solenoid operated mechanisms

6.7.1   Mechanisms for contactors

A principal characteristic of a contactor is the large number of mechanical operations, a million or more, that it is expected to perform. They are typically used for starting and stopping motors. Vacuum contactors have been successfully applied in mining and crane applications and a variety of industries including pulp and paper, chemical, and food processing. Contactors must handle the starting current of motors which is typically six times the full load current and persists, though declining, for some seconds with a high inertia load. Contactors do not normally interrupt fault current which is taken care of by a series-connected, fast-acting, current limiting fuses. Fast opening and closing is not required, so a relatively simple mechanism which closes in 100–350 ms and opens in 30–60 ms will suffice. This requirement is readily met by a solenoid-driven operator.

A DC coil pulls an armature, closing a magnetic circuit. The rotating action is communicated by simple axle and crank mechanism to the three poles of the contactor. The same action extends the opening springs so they are ready to power the next opening, and compresses the wipe springs of the individual interrupters.

Some contactors have a latch, the purpose of which is to hold the contactor closed without the need for continuous coil power. As mentioned, the contactor

Figure 6.21 *NEMA size-4 vacuum recloser*

(Courtesy of Westinghouse)

is closed (latched) by energising the main contactor coil. Once the contactor is latched the power to the main coil is removed by means of a late-opening, normally-closed auxiliary contact mounted on the contactor. To open (unlatch) the contactor, power is applied to the unlatch coil on the latch mechanism. Once the contactor is open the power to the unlatch coil is removed by means of a normally open auxiliary contact also mounted on the contactor.

Figure 6.22   Draw-out vacuum contactor
     (Courtesy of General Electric Co.)

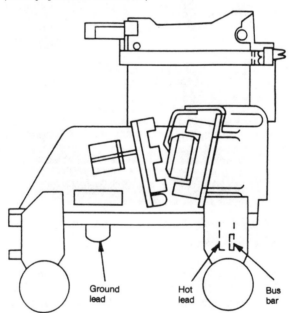

Figure 6.23   Draw-out vacuum contactor
     (Courtesy of General Electric Co.)

Vacuum contactors are available in a wide range of sizes and ratings. Figure 6.21 shows a very small contactor (NEMA size 4) used for motors rated 200V/40 HP to 575V/400 HP. The overall dimension of this device is about 15 cm. The interrupters for this contactor are illustrated on the left in Figure 5.5. In contrast, Figure 6.22 is a photograph of a draw-out vacuum contactor whose principal dimension is about 1 m. It can be used on motor at 2400V and 4800V with ratings of 1500 and 3000 HP, respectively. Note the vertically-mounted current limiting fuse of the first pole. The magnet, armature and opening springs can be seen in Figure 6.22 but the diagram of Figure 6.23 perhaps shows them more clearly. The three vacuum interrupters are mounted horizontally behind the barrier, bottom right. Details showing the drive can be seen in the cutaway of Figure 6.24.

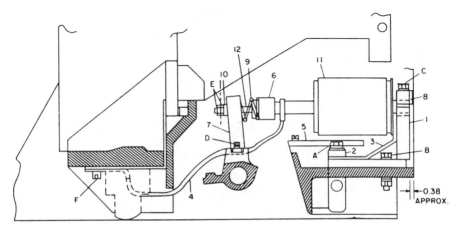

NOTE: DIMENSIONS IN INCHES

Figure 6.24   Cut-away showing vacuum interrupter and its mechanical connection to drive in draw-out contactor depicted in Figures 6.22 and 6.23

(*Courtesy of General Electric Co.*)

1 Interrupter support	*7 Moveable interrupter support*
2 Bus bar	*8 Interrupter clamp*
3 Bus strap	*9 Interrupter contact spring*
4 Shunt	*10 Brass sleeve*
5 Insulation shield	*11 Vacuum interrupter*
6 Insulator assembly	*12 Insulator shaft*

6.7.2   Mechanisms for switches

Switch mechanisms like those for contactors, are nearly always solenoid operated. Their stroke is frequently longer and they may actuate two or more

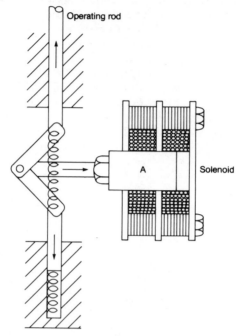

Figure 6.25   Simple solenoid operator for vacuum switch

interrupters in series. A very simple arrangement is that shown in Figure 6.25. When one of the coils of the solenoid is energised, the armature A moves to the right, causing the operating rod to move up to close the contacts (not shown).

The solenoid current is reduced to a low holding value. Subsequently, when the switch is to be opened, a bucking surge of current is applied to the other solenoid coil, which overcomes the constraint and drives A to the left to its original position.

A distribution switch with this style of mechanism is shown in Figure 6.26. This is a single-pole device but three can be ganged for three-phase operation, alternatively each can have its own independent mechanism.

The design shown in Figure 6.27 is used for three-phase switches. The interrupter columns (line-to-ground insulator and the interrupter itself) are bolted to the casting (shown cross hatched).

Two such locations are shown, above points D and E, the third pole would be at the edge of the diagram on the right. The mode of operation can be understood if one recognises that the toggle link rotates around A and is connected to the actuating bar at B, whereas the actuating bar moves in an arc around D and E. This switch, as illustrated, is open; the spring assembly connected between the toggle link and the control yoke holds the open stop against the bumper. When the control circuit energises the closing solenoid, the actuating pin rotates the control yoke clockwise until the toggle link is over centre. The spring assembly pulls the toggle link and the actuating arm in a

Figure 6.26 *Single-pole switch with solenoid operator*
(Courtesy of Joslyn Hi-Voltage Corp.)

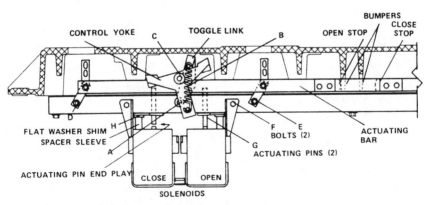

Figure 6.27 *Solenoid-operating mechanism for three-pole vacuum switch*
(Courtesy of Joslyn Hi-Voltage Corp.)

counter-clockwise arc until the close stop engages with the bumper. Vertical movement of the actuating bar raises the switch pull rods and closes their contacts.

When the opening solenoid is energised, the control yoke is rotated counter-clockwise until the toggle link is over centre and pulled clockwise by the spring assembly. This returns the actuating bar to the original position and opens the vacuum contacts.

Switches with this style of mechanism are found in Figure 7.4.

6.7.3   Solenoid design

We have written freely in the last sections of the use of solenoids, but said little about their design. This matter is addressed now, touching on general principles, force and motion characteristics, size and weight, etc. It is also shown how modern computational techniques have greatly improved design capability.

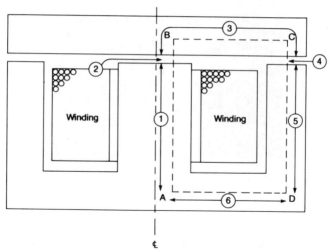

Figure 6.28   Rudimentary solenoid

Figure 6.28 shows the basic elements of a typical solenoid: the core or magnet, the winding, and the movable armature. The core and armature are made of laminated steel, so that when the coil is energised a magnetic field is set up with flux lines typified by the dotted line of Figure 6.28. Note that ①, ③, ⑤ and ⑥ of the magnetic path is in the steel, while parts ② and ④ are in the air gaps. Energising the coil causes the armature to be attracted to the core. The magnitude of the force between these two objects is determined by the magnetic flux density in the air gaps, which depends in turn on the reluctance R of the flux path and the magnetomotive force (MMF) produced by the coil. At each point in the magnetic circuit there is a magnetising force H, associated with the MMF through the magnetic circuit law

$$\int H_s ds = \sum i = MMF \text{ in the circuit} \qquad (6.7.1)$$

H is measured in ampere turns per metre, and at each point of the magnetic circuit the flux density is a function of H:

$$B = \mu H \text{ weber/m}^2 \qquad (6.7.2)$$

where μ is the permeability.

$$\phi = B \times \text{area} = \frac{MMF}{R} = \frac{V_A - V_B}{R} \qquad (6.7.3)$$

where $V_A - V_B$ is the difference magnetic potential between two points A and B. Making a few simplifying assumptions one can apply these equations to the magnetic circuit of Figure 6.28 to determine the current necessary for a given flux in the air gap. The change of magnetic potential is calculated round a closed circuit such as ABCD, approximately following a line of force. The circuit is divided into portions, here numbered ① through ⑥, each uniform in material and sensibly uniform in cross-section. The further calculation can be set forth in the form of Table 6.1 which is almost self-explanatory.

Table 6.1 *Compilation of ampere-turns for series parts of magnetic circuit*

Portion	Material	Area	$B = \Phi/A$	H	Length	$H\ell$
1	iron	A_1	B_1	(curve) H_1	ℓ_1	$H_1\ell_1$
2	air	A_2	B_2	(B_0/μ_0) H_2	ℓ_2	$H_2\ell_2$
3	iron	A_3	B_3	(curve) H_3	ℓ_3	$H_3\ell_3$
etc.	---	---	---	---	---	---

The flux in portions ① and ② divides into two equal parts, one of which traverses ③, ④, ⑤, ⑥. This may be allowed for by assuming a uniform circulating flux, but doubling the areas ③ to ⑥. The flux densities B_1, B_2, etc. having been thus found, the corresponding values of H are obtained from the magnetisation curve appropriate to the material being used, and from $B = \mu_0 H$ (eqn 6.7.2) in the air gaps. Since H lies everywhere close to the direction of the circuit, the sum of the last column $\Sigma H l$ is a good approximation to $\int H_s \, ds$, and so approximately equal to the ampere-turns required to set up the flux Φ.

The flux needed depends on the force required. To this end use Equation 6.2.5:

$$F = \frac{B^2}{2\mu_0} A \qquad (6.2.5)$$

Combining with Equation 6.7.2

$$F = \frac{H \times B}{2} \cdot A \qquad (6.7.4)$$

(which is in some ways more fundamental). As a rough approximation one can ignore the ampere–turns required by the steel and think only of those required by the air gap. Suppose the total length of these is 2 mm and $B = 1$ weber/m^2. To obtain a force of 60 newtons (about 25 lb) with an area of 10 cm^2 ($= 10^{-3}$m^2), then requires according to Equation 6.7.4

$$H = \frac{2 \times 60}{10^{-3}} = 1.2 \times 10^5 \, \text{ampere} - \text{turns per metre}$$

Thus, for 2 mm, one needs 240 ampere–turns. If the steel position of the magnetic circuit is 80 times as long and the permeability is 200 turns as great, the steel would require $240 \times 80/200 = 96$ ampere–turns, making a total of 336 ampere–turns. A coil with 100 turns would require 3.36 A.

This is a very crude calculation for a static condition, but it gives a first cut idea of what is needed. More refined figures can be obtained from such sources as Reference 11. In fact, the energising of a solenoid is not a static situation. In the first place, the current takes time to rise because of the inductance of the coil.

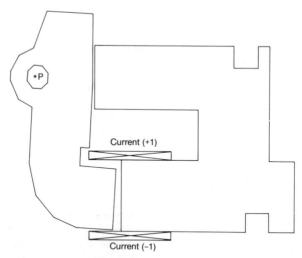

Figure 6.29   Cross-section of printer actuator [12]

> *Number of turns on coil* $= 196$
> *Hammer inertia* $= 1 \cdot 966 \times 10^{-6}$ kg
> $R_{ext} = 0 \Omega$
> $R_{DC} = 3 \cdot 89 \, \Omega$
> $L_{ext} = 90 \, \mu\text{H}$

Secondly, as the armature commences to move and the air gaps shorten, the force and the inductance both increase. We are dealing with a system in which current, magnetic field and motion are all coupled.

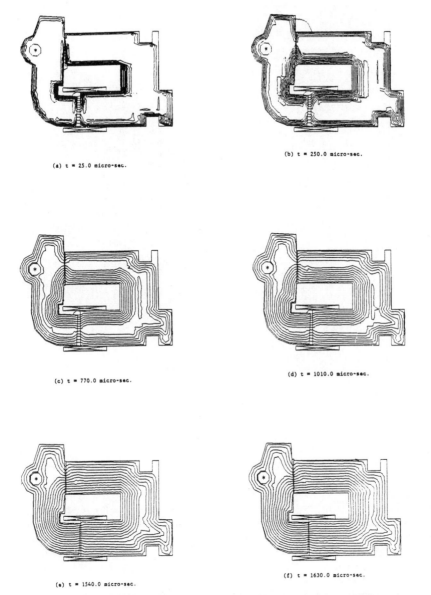

Figure 6.30 *Printer actuator: equipootential plots and hammer position at different instants*
[12]

Modern finite-element methods of analysis provide solutions to problems of this
kind. An example by Salon *et al.* [12] is cited by way of illustration. It deals with
a printer actuator and has much in common with a solenoid.

An impact printer mechanism is an electromechanical device which works as
follows (Figure 6.29). A voltage pulse is applied to a series coil. The current in
the coil produces a magnetic field which attracts the hammer which is made of

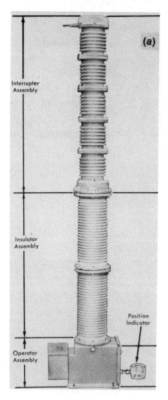

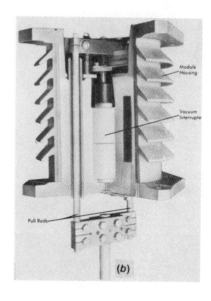

Figure 6.31(a)   Vacuum switch with five series interrupters
(b)   Cutaway showing how individual interrupters are mechanically coupled
(Courtesy of Joslyn Hi-Voltage Corp.)

magnetic material. There is an opposing force produced by a spring which will eventually return the hammer to its initial position. Thus, it is much like solenoids used for operating switching devices. Eddy currents in the magnetic steel delay the penetrated flux and the production of magnetic force. The problem is truly a coupled one: to find the current in the coil one must know the inductance which is a function of the position and material properties; to find the field and material property one must know the current and position; and to find the position one must know the forces which depend on the fields. The entire system must therefore be solved simultaneously.

From the initial state of no fields, no current, and no motion, the coil is excited with a pulse of voltage. The sequence of equipotential plots in Figure 6.30 shows the flux penetration (diffusion) into the steel. In the later pictures the gap is considerably reduced and almost disappears.

The printer actuator involves rotary motion rather than translation, which must be reflected in the global equations. Force is replaced by torque, mass by moment of inertia, velocity by angular velocity, and displacement by angular position. Salon *et al.* [12] show plots of torque, angular displacement and coil current for the actuator example.

Large solenoids take quite a surge of current when they are energised, it can be as much as that taken by a 1000 W incandescent light bulb. The auxiliary power source, whether AC or DC, must be stiff enough to accommodate the needs of the solenoids.

6.8 Mechanism for series operation of interrupters

High-voltage applications require the series connection of vacuum interrupters, whose contacts must close and open in unison. Failure to do so can cause problems, as noted in Section 4.8. This situation is best insured by having a relatively rigid coupling between the interrupters. Figure 6.31 shows how this can be accomplished. Figure 6.31(a) shows the assembled switch with five interrupters in series and appropriate insulation to ground, while Figure 6.3(b) shows a section through the bottom interrupter. Note the two pull rods which at the top connect to the cross clamp which drives the moving contact and at the bottom is joined to the single operating rod which passes through insulator assembly to the operating mechanism at the base. The pull rods continue upward to the top of the interrupter stack. Cross clamps connect them to the remaining interrupters.

6.9 References

1 HOLM, R.: 'Electric contacts' (Springer–Verlag, Berlin, 1967, 4th edn.)
2 BARKAN, P., and TOUHY, E.J.: 'A contact resistance theory for rough hemispherical silver contacts in air and in vacuum,' *IEEE Trans.*, 1965, **PAS–84**, pp. 1132–1144
3 RICH, J.A., FARRALL, G.A., IMAM, I., and SOFIANEK, J.C.: 'Development of a high-power vacuum interrupter', EPRI report EL-1895, Palo Alto, CA, USA, 1981
4 ANDERSON, J.M., and CARROLL, J.J.: 'Ability of a vacuum interrupter as the basic element in HVDC breakers', *IEEE Trans.*, 1978, **PAS-97**
5 Standard for Industrial Control Equipment, UL 507, 16th edn, 1993
6 BARKAN, P. 'A study of the contact bounce phenomena', *IEEE Trans*, 1967, **PAS–86**, pp. 231–240
7 SHAMES, I.H.: 'Engineering mechanics, vol. II, dynamics' (Prentice-Hall, Englewood Cliffs, NJ, USA, 2nd edn.) p. 447 et seq.
8 BARKAN, P., LAFFERTY, J.M., LEE, T.H., and TALENTO, J.L.: 'Development of contact materials for vacuum interrupters', *IEEE Trans.*, 1971, **PAS–90**, pp. 350–360
9 BARKAN, P., and McGARRITY, R.V.: 'A spring-actuated, cam-follower system: design theory and experimental results', *Trans. ASME, Series B J. Eng.*, 1965, **87,** pp. 279–286
10 BARKAN, P.: 'Dynamics of high capacity outdoor oil breakers', *Trans. AIEE*, 1955, **74**, pp. 671–676
11 McLYMAN, W.I.: 'Transformer and inductor design handbook' (Marcel Dekker, New York, 1988, 2nd edn.)
12 SALON, S.J., DeBORTOLI, M.J., and PALMA, R.: 'Coupling of transient fields, circuits and motion using finite element analysis', *J. Electromagn. Waves Appl.*, 1990, **4,** pp. 1077–1106

Chapter 7

Design of vacuum switchgear 3: design for versatility

7.1 Basic notions

The fact that a vacuum interrupter is a sealed, self-contained unit has some profound implications, for it means that it can be 'plugged in', or at least adapted, to a wide variety of breakers with little more difficulty than screwing in a light bulb. Great advantage has been taken of this in recent years. We are reminded of the situation that developed around 1970, which is described in Section 1.7, when growth of the market in vacuum power switching devices stalled after a promising start. Momentum was restored when the flexibility of the interrupter was recognised. The focus then was on the *inside* of the interrupter, where changes could be made which affected the rating of the interrupter without significantly changing it externally. Of course, this option is still exercised, but here the concern is with relatively simple changes in the remainder of the breaker package which will capitalise on the interrupter capabilities to produce a whole range of breaker ratings, both current and voltage. This is what is dubbed 'design for versatility'. It is also an important component of design for manufacturability.

The profile shown in Figure 7.1(a) contains three basic elements: the operating mechanism A, the interrupter C, and the support insulators B. This is a side view, so one can see only one pole, the other two are of course in line behind. For the sake of simplicity the stabs to the busbars, the rollers and other components have been omitted.

Figure 7.1(a) can be described as the base case, perhaps a 1200A, 12 kV breaker with an interrupting capability of 20 kA, like that shown in Figure 7.1(b). What is about to be examined is how this base case is modified to meet requirements of higher voltage, higher continuous current and higher short-circuit interrupting capabilities demanded by other applications.

The same philosophy is being applied to switches as well as circuit breakers. Therefore some of these designs are examined at the same time.

7.2 Designing vacuum switchgear for a range of ratings

7.2.1 Adapting to higher voltages

Higher voltage applications call for higher voltage interrupters, which in general are longer. This is mainly to increase the external flashover voltage. Figure 7.2

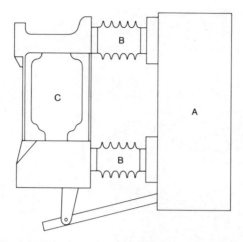

Figure 7.1(a)   Silhouette of basic vacuum circuit breaker

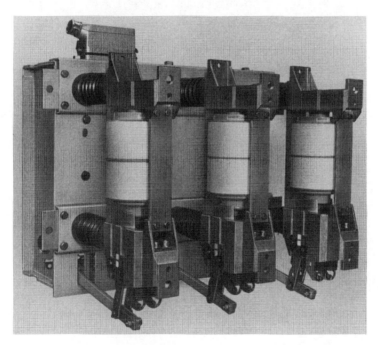

Figure 7.1(b)   Example of basic breaker
(Courtesy of Siemens AG)

shows how this has been very simply accomplished by one manufacturer. The support insulators, B in Figure 7.1(a), have been lengthened, and the top mounting channel has been moved upwards to accommodate the longer interrupter. Some manufacturers introduce insulating phase barriers between

Figure 7.2   Vacuum circuit breaker for 24 kV

(*Courtesy of Siemens AG*)

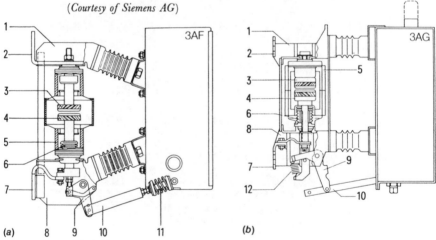

Figure 7.3   Outlines of standard breaker (a) and higher voltage version (b) with canted support insulators

(*Courtesy of Siemens AG*)

1 Upper interrupter support	7 Bottom terminal
2 Top terminal	8 Lower interrupter support
3 Fixed contact	9 Angled lever
4 Moving contact	10 Insulated coupler
5 Interrupter housing	11 Contact pressure spring
6 Metal bellows spring	12 Contact pressure release spring

the poles so as to guard against phase-to-phase sparkover. The mechanism remains unchanged.

An alternative favoured for higher current applications, is to cant the support insulators at an angle, as shown in Figure 7.3 which also shows the base case breaker for comparison. Once again, the mechanism is essentially the same.

A family of vacuum switches which follows a similar modular approach is illustrated in Figure 7.4.

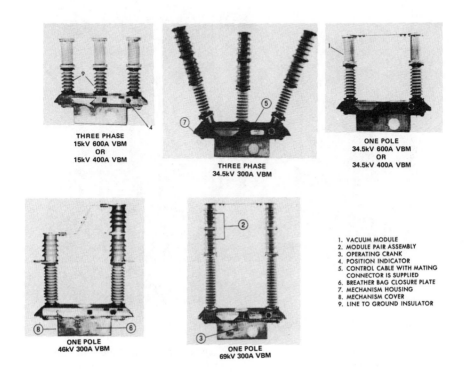

THREE PHASE
15kV 600A VBM
OR
15kV 400A VBM

THREE PHASE
34.5kV 300A VBM

ONE POLE
34.5kV 600A VBM
OR
34.5kV 400A VBM

ONE POLE
46kV 300A VBM

ONE POLE
69kV 300A VBM

1. VACUUM MODULE
2. MODULE PAIR ASSEMBLY
3. OPERATING CRANK
4. POSITION INDICATOR
5. CONTROL CABLE WITH MATING
   CONNECTOR IS SUPPLIED
6. BREATHER BAG CLOSURE PLATE
7. MECHANISM HOUSING
8. MECHANISM COVER
9. LINE TO GROUND INSULATOR

Figure 7.4   Modular vacuum switches for voltage ratings from 15 to 69 kV

(Courtesy of Joslyn Hi-Voltage Corp.)

There are a number of noteworthy points to be mentioned. First, all five equipments have essentially identical mechanisms. However, in the first two models the mechanism drives the three poles of a three-phase switch, whereas the remaining three examples are single-pole devices, three would be required for a three-phase unit. Incidentally, the mechanism used in all these switches is that shown in Figure 6.27.

Secondly, observe how devices of increasing voltage rating are obtained by connecting two or more identical interrupters in series. The corresponding increase in the insulation to ground is achieved by the length of the post insulators on which the interrupters are mounted. These are hollow of course, to allow the insulated operating rods to pass up them from the mechanism to the interrupters.

7.2.2   Adapting to higher load current

The problem here is thermal, more losses are being generated with increased current, so more energy must be extracted and transferred to the surroundings. This is achieved by the use of fairly massive heat exchangers which take the form of aluminum castings, with heat transfer fins as are shown in Figure 7.5.

Figure 7.5   Vacuum circuit breaker for high continuous current (4000 A) and high interrupting current

(Courtesy of Siemens AG)

This same photograph appears in Figure 5.21 in Section 5.7.3 where the thermal importance of the flexible braids is discussed.

Yet another excellent example of design for versatility, involving high continuous current, is to be found at the end of Section 10.3.2.

7.2.3   Adapting to higher short-circuit current

Concern here is with the high electromagnetic forces that high fault currents can produce. Again, Figure 7.5 is relevant. The mounting steel plate which forms the back of the mechanism has been replaced by a casting, which is ribbed for added strength. Observe how the bracing straps in parallel with the interrupters have been made considerably heavier. But note also how the basic design remains intact.

Of course, the heavier contacts have more inertia and require more force for

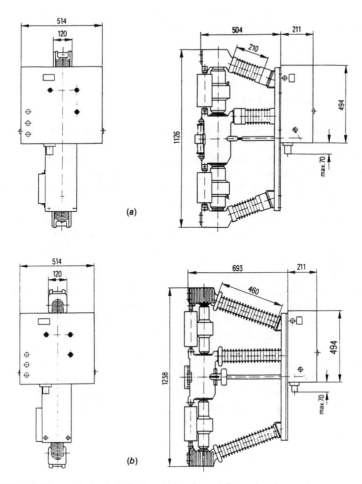

Figure 7.6 Outlines of single-pole circuit breakers (dimension in mm)

    (a) 17.5kV applications
    (b) 27 kV applications

    (Courtesy of Siemens AG)

their acceleration. Likewise the higher currents increase the popping force (Section 6.2). Both effects are readily countered in the choice of opening spring and wipe springs of higher gradient.

7.3 Single pole breakers

The ideas discussed in Section 7.2 can be readily applied to single-pole circuit breakers, an example used successfully for railroad applications appears in Figure 4.37. This is a higher voltage device and therefore employs two interrupters in series. Observe the progression from 17.5 to 27 kV indicated by the dimensional outlines of Figure 7.6. Applied indoors these are mounted on withdrawable trucks; for outdoor applications they can be installed in sheet metal, weatherproof, prefabricated substations. Once again the mechanisms are identical.

7.4 The OEM market

It will be apparent from Chapter 9, if it is not already, that the manufacture of vacuum interrupters is not trivial; it requires considerable expertise, specialised equipment and therefore considerable investment. This provides opportunities for those who are prepared to get into the act, to cater to the demands of a thriving Other Electrical Manufacturer (OEM) market. This market comprises a large number of companies, many of them quite small, which do not have the resources and/or the desire to make interrupters, but which can turn out an acceptable switchgear product with interrupters made by somebody else. Such companies are to be found in the developing as well as the over developed countries of the world.

There are many variations in requirements of OEMs but most of them are relatively minor, such as the number and location of mounting studs, the diameter of the connecting shank of the moving contact, the threads used (English or metric), and so on. These are relatively easy for the interrupter manufacturer to accommodate. The requirements of different clients can be stored on tape if the manufacturer uses numerically controlled (NC) machines and can be called up as needed. This is all possible because the interrupter is a sealed, self-contained unit, which gives the OEM the opportunity to adapt interrupters to his designs. Attributes such as the fact that the interrupter can be mounted in any position are particularly useful.

Chapter 8

Design of vacuum switchgear 4: packaging

8.1 Introduction

Tradition dies hard in the electrical equipment industry, designs for new products evolve relatively slowly. This is understandable because of the heavy emphasis on reliability. It is particularly true for switchgear, whose principal function in many instances is to protect other equipment. 'If it works, don't fix it,' has surely been the philosophy of many utility engineers. The advent of a new technology rarely results in a rapid change of direction. The replacement of mercury arc rectifiers by solid-state thyristors is the only exception that readily comes to mind. Some of the early SF_6 circuit breakers look remarkably like the oil breakers of that period. The new was adapted to the old, like the early automobiles that had their front seats high to see over the horse's back. As noted in Section 1.7, it took quite a time for vacuum to penetrate a conservative market, and by the same token it took time for manufacturers to recognize the special attributes of vacuum and adjust their designs accordingly. Nowhere is this more evident than in packaging.

The specific characteristic of vacuum that lends itself to versatile packaging is the completely self-contained character of the interrupters. An individual can pick up and hold an interrupter of the highest rating; in most instances one can embrace the interrupters of all three poles, like coconuts or basketballs. Their physical orientation, horizontal or vertical, is of no concern. Having said this, one must recognise that there is much more to a piece of switchgear than the interrupters. In metal clad circuit breakers, for example, there are many components, CTs, PTs buswork, auxiliary contacts, interlocks, etc. that are required, regardless of the switching technology. All these components require space, and more important, clearance between live parts, which is even more space, in most instances, and tend to make innovation a challenge. As in other chapters, the focus tends to be on those aspects of packaging that are unique or at least special to vacuum.

Section 5.7.1, particularly in Figure 5.20, explained how the size of vacuum interrupters has been reduced as their development has progressed. As the evolution of packaging is traced, observe that similar forces have been at work with the aim of reducing the size of the overall equipment and the space it occupies.

8.2   Air-insulated equipment

8.2.1   Traditional metal-clad gear

There is a long tradition of medium voltage metal clad switchgear for utility, industrial and commercial purposes. The equipment is disposed within steel cubicles. One cubicle will contain the breaker, others will enclose buswork or cable terminations. A number of such units, placed side by side, constitutes a 'line-up'. It forms a grounded metal structure that can be safely approached from front, back, or sides.

This tradition was continued when vacuum switching technology was introduced and the existing circuit breakers were replaced by vacuum circuit breakers. In most instances the breakers were mounted on wheels or rollers so that, when in the open position, the cubicle door could be opened and the breaker removed for servicing. In the majority of cases the breakers were, and still are, 'horizontal drawout,' that is to say, they are inserted horizontally or withdrawn horizontally, into or from their cubicles. In a few designs the breaker is rolled into the cubicle horizontally, and then jacked vertically into position. In either event, stabs on the breaker engaged with the live bus, placing the breaker in service. Such spring-loaded stabs are clearly evident in Figure 6.6. When the breaker is withdrawn, masks or shutters move automatically over the ports to preclude access from the breaker compartment at the front to the live bus compartment behind. This renders disconnect switches on either side of the breaker unnecessary.

Some early metal clad vacuum switchgear in the UK was of the fixed variety, that is to say, the breaker part could not be removed but was integral with the buswork. However, the trend worldwide, appears to be towards the withdrawable type.

8.2.2   Multiple stacking of circuit breakers

Vacuum breakers introduced into metal-clad switchgear during the 1960–1970s were frequently smaller than the equipment they were replacing. This was surely true for the air magnetic breakers in the USA. It is not surprising, therefore, that there was pressure from users to make use of the available extra space. Initially, the pressure came from commercial and industrial users, who were anxious to save space in industrial and commercial buildings. The design response of the manufacturers was to stack units one above the other, so that the space originally assigned to one breaker was essentially occupied by two. This two-tier, or two-high arrangement is now almost universally produced for medium-voltage line-ups.

Figure 8.1 shows an example of a two-high vacuum switchgear equipment. Apparatus of this kind provide centralised control of medium-voltage generators, motors, feeder circuits, and transmission/distribution lines for industrial, commercial and utility installation. Available configurations for the upper and lower tiers are shown in Figure 8.2. This disposition of components within the cubicles can be appreciated from Figure 8.3 which depicts sectional views of typical equipment; the dimensions are in inches.

Figure 8.1   Two-high medium-voltage vacuum switchgear installation
   (Courtesy of Westinghouse Corp.)

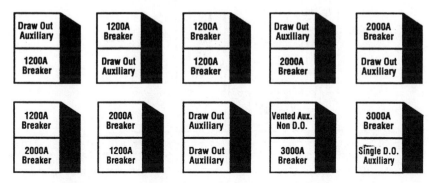

Figure 8.2   Optional arrangements for two-high vacuum switchgear
   (Courtesy of Westinghouse Corp.)

As shown, Figure 8.1 is an indoor installation, however, it can be adapted for outdoor use by placing it in an appropriate weatherproof housing. Several different arrangements are available. It can be a single line-up with, or without a protected aisle, or it can be a double line up, with the equipments facing each other across a common aisle. These alternatives are illustrated in Figure 8.4.

Chapter 7 is devoted to design for versatility. The preceding remarks on equipment arrangement indicate that it is not only in the breaker itself that versatile design is possible, but in the entire vacuum installation.

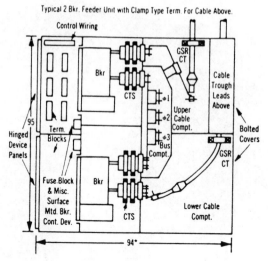

Figure 8.3   Sectional views of typical medium voltage two-high vacuum switchgear units
(Courtesy of General Electric Co.)

8.2.3   Air-insulated distribution switchgear

The first vacuum switching devices on utility systems, were air-insulated [1]; they were used for load and capacitor bank switching. The first vacuum power circuit breaker, a recloser [2], was also air-insulated. Introduced in 1962, many are still in service today. Air-insulated switchgear continues to hold a place in the array of vacuum switching equipment to be found on power systems.

Figure 8.5 shows a contemporary distribution breaker rated 15.5 kV and either 1200 or 2000A carrying; the symmetrical interrupting capacity is 20 kA.

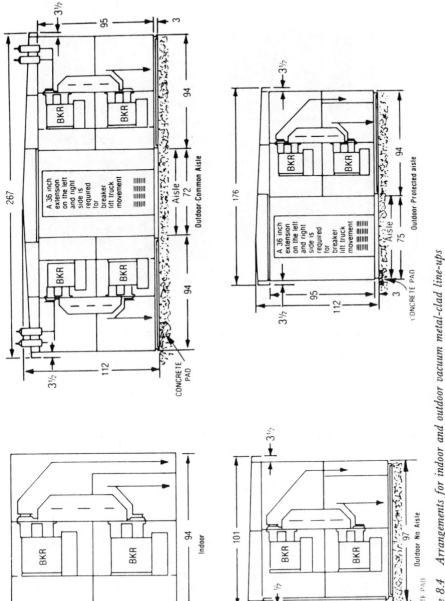

Figure 8.4   Arrangements for indoor and outdoor vacuum metal-clad line-ups

(Courtesy of General Electric Co.)

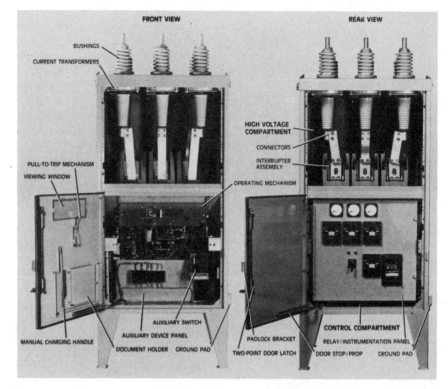

Figure 8.5   Contemporary, air-insulated, vacuum distribution breaker
(Courtesy of General Electric Co.)

This device, which has reclosing capability, occupies an area somewhat less than 4 ft square and stands rather more than 8 ft tall. The modular breaker design consists of three pole assemblies and the breaker mechanism. An entire module can be removed with a minimum of effort, thereby simplifying maintenance.

8.3   Foam-insulated switchgear

Some equipments which were air-insulated are now insulated with polyurethane foam. Encapsulating the interrupter in foam allows shorter interrupters to be used by increasing the external flashover. Foam also impedes the ingress of moisture and is not a good surface for water condensation, both of which help electrically. A closed foam, filled with freon, is presently used. It is likely that a substitute for freon will have to be found for environmental reasons. Figure 8.6 shows a cutaway of a 15 kV switch which embodies this feature. It is typically used in capacitor bank and reactor switching. A 34.5 kV distribution sectionaliser is shown in Figure 8.7. Each pole has two principal elements, the upper one is the interrupter; this interrupter has exterior elastomeric weather sheds made of ethylene propylene rubber with hydrated alumina as a filler.

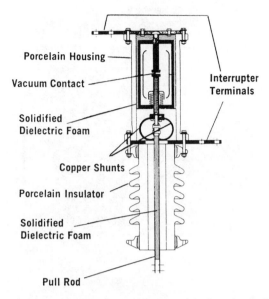

Porcelain Housing

Vacuum Contact

Interrupter Terminals

Solidified Dielectric Foam

Copper Shunts

Porcelain Insulator

Solidified Dielectric Foam

Pull Rod

Figure 8.6   15 kV vacuum switch utilising solid-dielectric foam insulation
(Courtesy of Joslyn Hi-Voltage Corp.)

Figure 8.7   34.5 kV, 600A distribution sectionaliser
(Courtesy of Joslyn Hi-Voltage Corp.)

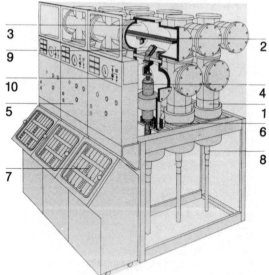

Figure 8.8   Vacuum switchgear for gas-insulated substation

(Courtesy of Siemens AG)

1 *Aluminum casting*   6 *Current and potential*
2 *Station bus transformers* 7 *Secondary arrangements*
3 *Insulation*   4 *Vacuum interrupters*
4 *Vacuum interrupters*   9 *Operator for disconnect switch*
5 *Control cabinet*

8.4 Gas-insulated equipment

8.4.1 Vacuum switchgear for gas-insulated substations

Gas-insulated switchgear was developed during the 1960s as a component for gas-insulated substations (GIS). At the 1972 CIGRE meeting, no fewer than seven reports were presented and discussed in the group 23 (substations) session [3]. This is an expensive technology, but is justified in many situations by the enormous saving in space that can be achieved in comparison with air insulated stations. Moreover, it was believed that this development would lead to greater reliability.

Gas-blast circuit breakers, disconnects and grounding switches were originally used for these applications and SF_6 technology continues to be used extensively today. The components are housed in tubular metal enclosures which are themselves joined by gas-tight seals to form a totally enclosed system. The whole is filled with SF_6 to approximately 2 atmospheres pressure. It became clear in due time that, because of their compact design and small mechanism, vacuum interrupters would be ideal candidates for GIS at the lower voltages. Figure 8.8(*a*) shows an example of gas insulated vacuum switchgear for a capacitor substation; Figure 8.8(*b*) is a cutaway of the same equipment which permits the various components to be identified. By way of calibration, the following numbers give some idea of the scale: overall height $= 2.1$ m, depth $= 1.51$ m, width of each section $= 0.60$ m.

8.4.2 Gas-insulated distribution switchgear using vaccum

Two other examples of gas insulated vacuum switchgear are to be seen in Figs. 8.9 and 8.10. The first, depicted in Figures 8.9(*a*) and (*b*) is a pole-mounted auto recloser.

It will be seen from the cutaway that each interrupter is housed within the upper part of the ethylene propylene diene monomer (EPDM) rubber housing which is filled with SF_6 at half bar gauge pressure. The dielectric filling is used only to provide the impulse voltage withstand capability. Full power frequency withstand is achieved with only air filling.

The EPDM rubber housing is molded to incorporate weather protection sheds and also accommodate the protection current transformer. Only static seals are used to contain the filling medium.

The operating mechanism, which is enclosed within the main housing, comprises a magnetic actuator which drives all three phases of the recloser. The principle of the actuator is that it has only one moving component, the armature, which is latched by means of a permanent magnet at either end of its travel. The neodymium iron boron magnet is well known for its ability to operate at high temperature for long periods without loss of magnetism.

The second example (Figure 8.10) is also a recloser. The three poles and the mechanism are mounted within the grounded steel tank, which is filled with SF_6. This is a low-pressure vessel that does not require special shipping and handling. What makes this design unusual is that, after assembly, the tank is welded closed. This clearly demands a high degree of reliability. Such a design can be described as maintenance-free, or unmaintainable, depending on one's point of view. The

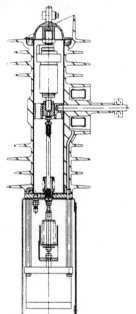

Figure 8.9   Pole-mounted auto recloser
(Courtesy of GEC Alsthom)

Figure 8.10   Completely sealed gas-insulated recloser

(Courtesy of Joslyn Hi-Voltage Corp)

15 kV, three-phase unit is compact (approx. 33 in long), and weighs only 130 lb
(59 kg).

8.5   Vacuum switchgear with composite solid/gas
insulation systems

The benefits of dielectric coatings on conductors to improve the electrical
breakdown strength of a gas system has been known for some time. Designs of
gas-insulated current transformers for airblast circuit breakers capitalised on this
technique in the late 1960s. Much more recently composite insulation systems of
this kind have been adopted for a new style of vacuum switchgear, cubicle gas-
insulated switchgear (C-GIS).

Simply placing a dielectric barrier in the gas space between a pair of electrodes
may not improve the breakdown strength of the gap. Indeed, it could decrease
the breakdown voltage because the higher permittivity of the dielectric increases
the field strength in the gas for a given applied voltage. However, coating the
electrodes with a skin of solid dielectric can effect a considerable improvement of
the gap breakdown properties, especially where the field is very nonuniform.
Yoshida *et al.* [4] show that under certain circumstances an improvement of close
to 100% is achievable.

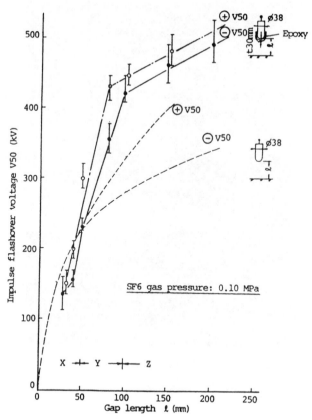

Figure 8.11   Impulse characteristic of gap with insulation layer [5]

The factors contributing to the increase in breakdown voltage include suppression of electron emission from the electrode by the insulating covering and suppression of negative ions by the lowered field intensity.

Figure 8.11, from a paper by Masaki *et al.* [5], shows the effectiveness of an insulating layer with respect to the impulse breakdown characteristic of a non-uniform vacuum gap. Comparing the covered electrode with the bare electrode, three regions can be identified, as indicated in the figure. These are

region X, where the gap length is short and the breakdown voltage is lower than the bare electrode for the reason already cited

region Y, when the insulating covering effects a considerable increase in the breakdown strength of the gap

region Z, where the gap length is long and the voltage characteristic saturates. However, the breakdown voltage is much higher than that of the bare electrode.

An example of the application of this technology to a piece of vacuum switchgear is shown in Figure 8.12(*a*), which shows two circuit breakers, one at either end of

(a)

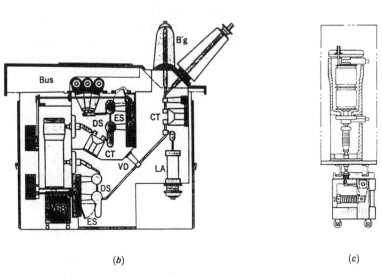

(b)   (c)

Figure 8.12   *Circuit breaker with solid/gas insulation system*

  (a) *Exterior view*
  (b) *Cutaway of interior*
  (c) *Vacuum element*

  (*Courtesy of Toshiba Corp.*)

the line-up with their associated transformer, etc. in the centre cubicles. A cutaway diagram of the circuit breaker cubicle is shown in Figure 8.12(*b*), while details of the circuit breaker itself can be seen in Figure 8.12(*c*). Equipments of this kind are available at ratings of 72 and 84 kV, with rated current of 800 and 1200 A, and interrupting currents of 20, 25 and 31.5 kA.

8.6 Oil-insulated vacuum switchgear

Tradition dies hard. Oil has been used satisfactory as an insulating medium for switchgear for very many years. It is not therefore surprising that when the switching technology changed to vacuum, oil was retained as the insolent. It continues to give good service in this role.

An example of an oil insulated recloser is to be seen in Figures 6.17 and 18 and oil is used in tap changers, as mentioned in Section 10.12. On one occasion, the author made a deliberate leak in a vacuum interrupter which was under oil, and then proceeded to open its contacts under load. A fault was created, but with the backup set with normal delay, there was no conflagration. Oil and vacuum are surely not incompatible.

8.7   Conversion kits

Soon after vacuum circuit breakers were accepted by users, kits for the conversion of other technologies to vacuum appeared on the market. In the US they were particularly directed at air-magnetic circuit breaker replacement. The more efficient vacuum breaker replaced the existing air-magnetic breaker without any change in the cubicle line-up. This saved the considerable expense of replacing the entire equipment. According to statistics kept by the National Electrical Manufacturers' Association (NEMA), more than 400,000 airmagnetic breakers were installed in the 5–15 kV range of medium voltage metal-clad switchgear from 1955–1985, in the US alone. This figure does not include other aging equipment, such as minimum-oil circuit breakers for fixed-mount substation applications. It is evident, therefore, that there is ample opportunity for business in this area, particularly in an environment in which life-extension of equipment has an important priority.

Figure 8.13 shows a vacuum retrofit breaker for 15 kV, 1000 MVA application. These breaker elements are supplied without the racking mechanism and carriage, primary contacts and secondary disconnects, since these are available in the original equipment.

Recently, Slade *et al.* [6] have described a retrofit for *oil circuit breakers*, but rather than just a replacement, it is intended to upgrade the device. The original bulk-oil breakers were rated 121–145 kV and 10 or 20 kA interrupting. With the oil interrupters exchanged for vacuum interrupters, the interrupting capability has been increased to 40 kA. How the modification has been brought about is evident from Figs. 8.14(*a*) and 8.14(*b*). It is claimed that the retrofit kit not only extends the breaker's life considerably, but it also provides a significant reduction of breaker maintenance.

Figure 8.13   Vacuum retrofit breaker for 15 kV, 1000 MVA applications

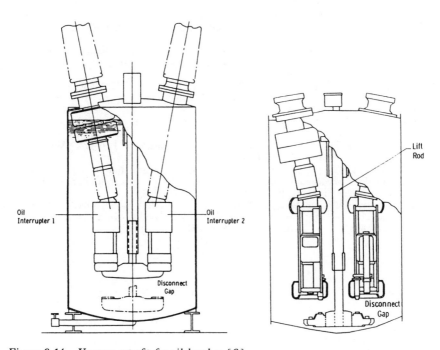

Figure 8.14   Vacuum retrofit for oil breaker [6]
(By permission of IEEE)

8.8   Hybrid circuit breakers

The idea of combining two power switching technologies to obtain the special advantages of both has been considered. Votta [7] presented a paper on an SF_6/ vacuum hybrid as long ago as 1976. More recently Natsui *et al.* [8] have also investigated the marrying of these technologies. The very rapid recovery of diffuse vacuum arcs has been noted at several points in this text. On the other hand, the vulnerability of SF_6 interrupters to fast-rising TRVs as in short line faults (Section 10.5.3) is well known. A series combination of vacuum and SF_6 interrupters might therefore take advantage of vacuum's dielectric capability during the initial recovery period, and the ability of SF_6 interrupters to hold off very high voltages over the longer term, which presents a problem to vacuum. Means must be found to assure the appropriate distribution of voltage between the vacuum and SF_6 breakers, throughout the entire recovery period.

Hybrid breakers have not appeared on the market at the time of writing, but may well be available in the future.

8.9   References

1 JENNINGS, J.E., SCHWAGER, A.C., and ROSS, H.C.: 'Vacuum switches for power systems', *Trans. AIEE*, 1956, **75**, pp. 462–468
2 STREATER, A.L., MILLER R.H., and SOFIANEK, G.C.: 'Heavy duty vacuum recloser', *Trans. AIEE*, 1962, **81**, pp. 356–363
3 Reports 23-01, 23-02, 23-03, 23-04, 23-08, 23-09 and 23-10, CIGRE, 1972, Paris, France
4 YOSHIDA, T., MIYAGAWA, M., OHSHIMA, T., MASAKI, N., and YANABU, S.: 'Increase of breakdown voltage due to composite insulation in SF_6 gas', *Electrifc. Eng. Japan*, 1991, **111**, pp. 36–46
5 MASAKI, N., OHSHIMA, T., YOSHIDA, T., and MATSUZAWA, R.: '72 kV C-GIS using composite insulation in SF_6 gas', Proceedings of 3rd IEE International Conference on Future trends in distribution switchgear', 1990, pp. 26–30
6 SLADE, P.G., VOSHAL, R.E., WAYLAND, P.O., BAMFORD, A.J., McCRACKEN, G.A., YECKLEY, R.N., and SPINDLE, H.V.: 'The development of a vacuum interrupter retrofit for the upgrading and life extension of 121–124 kV oil circuit breakers', *IEEE Trans.*, 1991, **PWRD**, pp. 1124–1131
7 VOTTA, G.A.: 'Novel concepts for interruption for distribution and transmission circuit breakers', Proceedings of symposium on New concepts in fault current limiters and power circuit breakers, EPRI Report El-276-sr, 1976
8 NATSUI, K., KURASAWA, Y., NAKAMATA, N., and YOSHIOKA, Y.: 'Voltage distribution characteristics of series connected SF_6 gas and vacuum interrupters after a large AC interruption', *IEEE Trans.*, 1988, **PD–3** , pp. 241–247

Chapter 9
Manufacture of vacuum switchgear

9.1 Organisation and general remarks

This chapter is concerned with the manufacture of vacuum switchgear, the emphasis being on *vacuum*. It concerns what is unique to vacuum; the fabrication of cubicles, the forming and assembly of busbars are not discussed, nor are provisions for CTs, PTs, relays, etc., all of which are common to every other switchgear technology. In practice, of course these items cannot be ignored, they are vital parts of the assembly line; but they are not addressed in this chapter.

As with most manufactured items, the process starts with the making of individual parts or components, which are then brought together into subassemblies; these in turn are assembled in a final product. Quality checks are made along the way and tests are performed on the final product. Steady flow is the objective, and most especially, steady flow with the minimum of inventory. In this regard a good figure of merit is the ratio of throughput to inventory. The recent trend, already referred to in the chapters on design, of reducing the number of parts and standardising on parts as far as possible, clearly helps in this regard.

In the manufacture of the interrupter there are some special considerations and constraints. For example, it is sound policy to get internal parts, once prepared, into their final vacuum environment as quickly as possible to avoid their contamination and degradation. Several processes - brazing, degassing - take a prescribed time because of the heating/cooling cycle. It is clearly desirable to accomplish these comfortably within a working shift. Throughput is then determined by batch size.

In the category of what is unique to vacuum, the interrupter is surely pre-eminent. Accordingly, this chapter begins with this component.

9.2 The vacuum interrupter

9.2.1 Contacts

Since contacts play a vital role in the performance of vacuum switching devices, their preparation is critical.

There are two ways of making the popular Cu/Cr contact material. The first begins by compressing the chromium powder into a cohesive mass, which is then sintered to form what might be described as a metallic sponge. The heating cycle is indicated in Figure 9.1. The procedure starts in vacuum but hydrogen is introduced as the process continues. A controlled flow of hydrogen is maintained

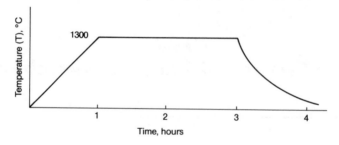

Figure 9.1   Thermal cycle for sintering chromium sponge in preparation of Cu/Cr contacts

during the high-temperature sinter. Dry nitrogen is introduced later to assist in the cooling period.

A second cycle is then initiated, during which the chromium is impregnated with copper. The copper is placed on top of the chromium sponge and subjected to the thermal profile shown in Figure 9.2. Once again the process starts in vacuum but hydrogen is added as the cycle proceeds. The hydrogen inhibits oxidation and can be easily removed afterwards by a vacuum bakeout.

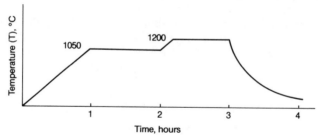

Figure 9.2   Thermal cycle for impregnating chromium sponge with copper in preparation of Cu/Cr contacts

The second method of preparing Cu/Cr contacts starts with the mixing of the two constituents in fine powder form; the mixture is then ground still further. Subsequently, it is compressed with considerable force into discs. A typical contact disc might be 6 cm in diameter and 1 cm thick; the force required for compression might be 90,000 kg. The discs are placed in sand inside a refractory container before being heated to the melting point of copper. The sand provides sufficient support to preserve the integrity of the discs, which are observed to shrink during the heating cycle.

A 60:40 mixture of Cu/Cr is usual for these contacts; a higher chromium contact (50:50) has improved dielectric performance but is more expensive. There has been considerable research over the years on Cu/Cr. It has revealed that superior performance is obtained if the grain size of the constituents is very small. A number of manufacturers assure this condition by repeated melting and freezing of the material. This procedure is carried out after the interrupters are completely assembled. Typically, the contacts are separated when they are

Figure 9.3   Induction furnace for preparation of copper/bismuth

   (*Courtesy of General Electric*)

carrying approximately 2,500 A, and the arc so produced is allowed to last for 60 ms or thereabouts. Fifty or sixty such treatments are carried out with a quite dramatic effect on the metallurgy of the surface material. It is found that the grain size is greatly reduced and this noticeably improves the dielectric strength.

   Many details on the metallurgy of Cu/Bi are to be found in Section 5.3.2. This material is usually prepared by induction heating, a furnace for this purpose is shown in Figure 9.3. Typically, the copper ingots are preheated in vacuum and then argon is introduced during the stirring cycle when the bismuth has been added.

   Considerable ingenuity is shown in the machining of the contacts so as to produce an effective product with minimum of effort. Figures 5.7(*a*) and 5.16 are examples.

9.2.2   Preparation of internal parts

All internal parts - contact shanks, shields, bellows, etc. - are rigorously cleaned before being assembled to assure that their surfaces are as free as possible of all contamination. Machined parts have an initial degreasing and then are washed with one or more agents. Deionised water is used subsequently, to be followed by ultrasonic cleaning in freon. The freon has been abandoned by one manufacturer because of environmental concerns and replaced by alcohol. However, because of the danger of explosion, the procedure is carried out in a pressure vessel. Figure 9.4 shows a typical cleaning line for component parts.

Figure 9.4   Cleaning line for component parts

(Courtesy of General Electric)

The surface condition of shields, *vis-a-vis* surface roughness, is also important from the viewpoint of dielectric breakdown. Rich *et al.* [1] performed a fairly exhaustive inquiry into this subject. They investigated components made from both bulk material and from spun material. After the surface contour had been made, whether by machining from the bulk metal or by spinning, the shaped parts underwent a variety of processing steps which included polishing with emery paper, chemical etching, and hydrogen or vacuum firing. It was found that each of these steps could alter the breakdown characteristics of the finished surface in vacuum. It is common practice to electropolish shields for 15 kV class and above.

9.2.3   Subassemblies

Figure 5.7(*a*) shows two typical subassembles, they comprise all the parts for both ends of an interrupter. One end has the moving contact, its shank, the bellows, the bellows shield and the end plate. The other end has the fixed contact and its support, together with the end plate. Some manufacturers carry out the brazing of the subassemblies in vacuum, some in a hydrogen atmosphere. In the

Figure 9.5   Vacuum furnace in clean room

(Courtesy of Calor–Emag/ABB)

last-mentioned case a tunnel kiln is often used. The components pass through in about five hours, but are at the brazing temperature for a few minutes only. One braze material used is copper–silver–palladium which melts in the range 850–900°C The braze is in the form of thin wavy washers.

If brazing is conducted in vacuum, the braze cycle may well be the principal degassing operation. Figures 9.5 and 9.6 show examples of the vacuum furnaces used.

The ceramic cylinders also form a subassembly in that the ends of the high alumna ceramic pieces must be metallised. This is a two-step procedure. A powder mixture of moly–manganese oxide and a binder is applied to the end of the cylinders which are then passed through a damp, reducing (H_2) atmosphere in a tunnel kiln. The kiln is divided into zones, the central zone assures that the ware reaches a temperature of 1400°C or thereabouts. The heating elements are molybdenum rods. The tunnel kiln cycle takes about six hours.

The second step is to build up the thickness of the metallised layer that is bonded to the ceramic. This can be done with braze materials or by plating.

In designs where there is a floating centre shield, the central flange of the shield may be sandwiched between two ceramic cylinders in a brazing operation. In some instances this is a separate operation, in others it is part of the final assembly. Where it is a two-step process, the first braze must have a higher melting temperature than the second.

Subassemblies are subjected to helium leak checking, a procedure which is usually automated. The detector should have a minimum sensitivity of 3 x 10^{-11}

Figure 9.6   Vacuum furnace being loaded with subassemblies
   (*Courtesy of General Electric Co.*)

standard cubic centimetre per second. A dozen or more units are tested simultaneously.

9.2.4   Final assembly

How final assembly is carried out depends on whether the subassemblies are welded or brazed together. Examples of edge welds will be found in Figure 5.19. This is now an automated process, performed by robots. It is conducted with an inert gas shield. I well remember when this was done laboriously (and frequently with much frustration) by hand.

The completed unit must now be baked out under vacuum before being pinched off. Bakeout temperature is 450° which is approached slowly. A useful procedure is to cool down after three hours or so and then perform a conditioning or spark cleaning test. This involves applying voltage across the contact gap from a relatively high impedance source. The voltage is raised to 80 kV for, say, 20 s, allowing breakdowns to occur as they will. The bakeout cycle is then repeated. The technique improves the interrupters dielectric capability.

Welded interrupters have a pinch-off tube through which they are evacuated. After bakeout the tube is sealed by pinching it between two steel rollers using an hydraulically powered tool. This essentially creates a cold weld along the line of the pinch. It is usual to dip this stumpy remnant of the pinch-off tube into solder. This reinforces the seal and at the same time masks the very sharp edge left by the pinch-off tool. Such a pinch-off tube is clearly visible in the photograph of the vintage interrupter in Figure 1.7.

When the subassemblies are joined together by brazing, the brazing cycle and the bakeout become the same operation. It is performed, of course, in a vacuum furnace. Small interrupters are preassembled for brazing inside a graphite jig. This requires more elaborate jigs but is easy to do inasmuch as parts are more or less dropped into place. The larger interrupters are fitted into harness-like jigs, which assure the concentricity of the parts. Rods with nuts hold the parts together axially. Differential thermal expansion of the rods and the envelope assure good pressure on the braze during the heating cycle.

Increasingly, the 'one shot braze' technique is being applied for vacuum interrupter assembly, especially for smaller interrupters. As the name implies, *all* the components are assembled at one time in one brazing process. The procedure is carried out in a vacuum furnace, which assures a good internal vacuum when the seals are made. This can be improved still further by a getter. One way of doing this is to have an almost complete copper ring within the envelope, with the gap in the ring spanned by a fine titanium wire. The wire is vaporised after assembly is completed by electromagnetic induction at a frequency of 10 kHz. The titanium acts as a getter and can reduce the degree of vacuum by two orders of magnitude.

9.2.5 Testing

As long as an interrupter remains on vacuum system it can be checked for leak tightness by a helium leak detector. The interrupter is bagged in helium, or a helium 'sniffer' is played along the joints, at the same time that a mass spectrometer, tuned to helium, searches for traces of the gas within the vacuum system. This is not possible once the interrupter is sealed. Completely assembled interrupters are routinely leak checked by a so-called 'magnetron' pressure measuring device.

The magnetron device applies simultaneously a high voltage between the open contacts and perpendicular magnetic field. Electrons produced by a small field-emission current are caused to gyrate, thereby taking a much longer path from cathode to anode. If residual gas is present, collision ionisation occurs and the current is amplified in what is sometimes called a *Penning discharge*. It is usual to observe the current collected by a biased shield, which is a measure of the background pressure. This is recorded for each interrupter. It is remeasured after a lapse of time, typically two weeks. A significant increase indicates a leaking interrupter. The procedure can be speeded up by storing the interrupters under pressure. If the ambient pressure is two atmospheres, the time between measurements can be reduced by a half.

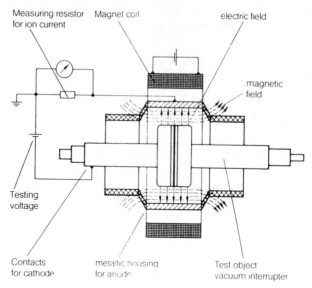

Figure 9.7   Schematic diagram of magnetron-type leak detector

(Courtesy of AEG)

A diagram showing the basic elements of a magnetron leak detector is shown in Figure 9.7. The interrupter in this illustration has an exposed central shield, making access to it easier.

If a vacuum interrupter has a leak, the pressure increase, Δp during a time interval Δt, depends on the quantity of gas Δm that penetrates the vacuum envelope during this time period. On the assumption of constant gas density on the outside, and no gas released from surfaces on the inside, the plot of pressure against time $p(t)$ gives a straight line when plotted on double logarithmic scale. The relationship can be stated

$$\ln(\Delta p V) = \ln L + \ln \Delta t \tag{9.2.1}$$

where V is the volume of the vacuum envelope, and L is the leak rate, i.e. the pressure change per second per litre. By recording two pressure readings two weeks apart, as described, it is possible to predict how long it will take for the internal pressure to reach some upper limit. Such a plot is presented in Figure 9.8. It shows that starting with an initial pressure of the order of 10^{-7} Pascal (7.5 x 10^{-10} torr), and experiencing a leak rate of 10^{-11} Pa /s, it would take about 10^9 seconds, or 31.7 years for the internal pressure to reach 10^{-2} Pa.

A test of this kind will not reveal an interrupter that has gone to atmospheric pressure. Defective devices can be identified by a simple hipot test when such a rare event occurs. Hipot tests are beneficial in any event, because of the conditioning they provide. Some manufacturers perform such tests under oil in a steel tank. This ensures that the integrity of the inside parts are stressed. The tank serves as a barrier against X–rays, which must always be a cause for concern whenever high voltage tests are being made.

Some manufacturers have a regimen of mechanical testing. Concern is for the

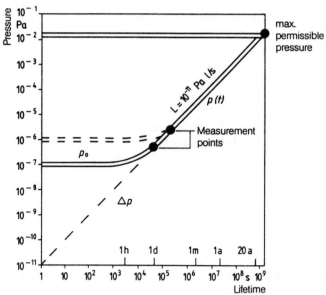

Figure 9.8 Pressure rise in vacuum interrupter as a consequence of a leak

(Courtesy of AEG)

ceramic-to-metal seals and their ability to withstand the pounding of mechanical operations. A small fraction of production is subjected to tensile testing to destruction. End plates should distort before joints part or ceramics fracture. Additional units are supported at the moving contact end and closed forcefully and repeatedly with the 'fixed' end unsupported. Again, this should not upset the joints.

9.2.6 Facilities

What sets vacuum apart from other power switching technologies is the absolutely pristine conditions which must be maintained where the interrupters are assembled. Sloppiness in this regard always has its price in poor dielectric performance. Clean rooms are usually maintained at two levels, the first is where parts are cleaned and washed and where subassemblies are put together. The second, with a higher degree of cleanliness, is for final assembly. Access to this is through trapped ports. Both kinds of clean room are maintained under positive pressure to obstruct the ingress of undesirable dust. The air is constantly circulated and filtered. Super clean rooms have a lobby. The internal doors to the cleanroom proper and the external doors to the remainder of the plant are interlocked so that both cannot be opened simultaneously. Ware is brought into the lobby on carts, then transferred to the cleanroom, which the carts themselves never enter. Airborne particulates are a cause of special concern. Thus, places where these are generated, such as contact fabrication facilities, are kept well separated from the cleanroom area and are maintained at slightly lower pressure by an external exhaust.

Extensive vacuum systems are, of course, essential. As the business has grown these have tended to become ever bigger. Interrupters are now processed in quite large batches. Cryogenic pumps are gaining popularity.

9.3   Breaker assembly

9.3.1   Breakers for metal-clad equipment

The topology of the assembly area is traditional. It is a long bay where the components move in from both sides and the flow is down the centre, along the axis of the area, proceeding to one end where the product is tested. What we have called 'design for versatility' pays off in this sequence. The commonality of many parts between different models reduces inventory and storage space. An operating mechanism may be standard for several ratings of breakers, except for the strength of the springs. These, then, are the only variant and can be colour coded to ensure that the right ones are fitted.

Following assembly, the breaker is subjected to as many as 300 mechanical operations, after which some 30 checks are made of such items as acceleration, bounce, and contact drop. All checking is performed automatically; data is recorded and printed out; all discrepancies are identified so that offending units can be withdrawn and fixed.

9.4   Reference

1 RICH, J.A., FARALL, G.A., IMAM, I., and SOFIANEK, J.C.: 'Development of high power vacuum interrupter', EPRI report EL-895, 1981

Chapter 10
Application of vacuum switchgear

10.1 Introduction

Vacuum switchgear covers a wide range, extending in interrupting capability from small contactors to powerful circuit breakers and in voltage from a few hundreds of volts to switches with series interrupters that operate at 230 kV. The purpose in this chapter is to learn how to select equipment for the vast number of duties to which it can be applied. This requires that we consider the stresses to which the switchgear is subjected in the various applications. In the process it will be discovered that there is a strong interaction between the switching device and the circuit in which it is operating.

10.2 Comments on standards

One cannot proceed very far with a discussion on switchgear application, still less on switchgear testing, without addressing the subject of standards. Standards serve several purposes: they define conditions for various applications, they propose methods for testing equipment to meet these conditions, and in some instances provide guide lines for applying the equipment. Thus, if a piece of switchgear has been designed to meet a particular set of standards, it is warranted to pass the tests designated by those standards. Standards also prescribe preferred ratings for equipment so as to avoid unnecessary proliferation of ratings.

The need for standards arose when manufacturers began to offer products for different applications. Agreement had to be reached between manufacturer and user concerning the stresses imposed by the applications and how to verify that the product could withstand the stresses. Standards organisations were set up and grew in different countries. In the US these were the American National Standards Institute (ANSI), the Institute of Electrical and Electronic Engineers (IEEE) and the National Electrical Manufacturers Association (NEMA). Similar organisations exist in other countries, in Europe, the UK and Japan, for example.

Development of international markets and the shift to a global economy necessitated international agreement on standards. This role has been met by the International Electrotechnical Commission (IEC). The establishment of standards often requires much technical investigation and data collection. IEC has been assisted in this over the years by the study committees and working groups of the Conference International des Grandes Réseaux Electriques (CIGRE). Of special relevance in the present context is the committee for switching equipment, which is CIGRE SC13. Congrès International des Réseaux Electriques de Distribution (CIRED) also addresses such matters at the distribution level.

The process of developing a standard can be laborious and time consuming, involving as it does the reaching of a consensus between the committee members who represent different constituencies, often with conflicting interests. It is not surprising, therefore, to find statements such as '. . . to be agreed upon by the user and the manufacturer', recurring in the document when it is ultimately published.

As new problems arise in the power system, new standards must be developed, or old ones modified, to meet the new situation. An example was the incidence of breaker failures in the 1950s and 1960s due to stresses produced by short line faults, that is faults one or two kilometres distance from the breaker along a transmission line. Travelling waves between the fault and the breaker cause a particularly fast-rising TRV [1] which many breakers could not sustain. Once the problem was analysed and understood, standards were written to define the phenomenon and testing methods were devised to determine the ability of a breaker to handle such an event. This information is incorporated in the current document 56 IEC 1987 in clauses 4.105 'Rated characteristics for short-line faults', 6.109 'Short-line fault tests', and Appendix AA, 'Calculation of transient recovery voltages of short-line faults from rated characteristics'.

Standards are voluminous documents. For example, document 56 IEC 1987 contains well over 300 pages. It is divided into two chapters: I Service Conditions, Definitions, Rating, Construction and Design, and II Tests, Selection, Orders and Installation, and eight appendices. This treatise concentrates on those points of difference between various standards addressing the same subject, and any special implications there may be with respect to vacuum. Regarding the first point, differences are to be expected because the standards have developed from different roots in different countries. However, these differences are becoming less with succeeding revisions and an earnest effort to move towards uniformity on the part of the committees that frame the documents. In the forward of 56 IEC 1987 the following statement appears: 'In order to promote international unification, the IEC expresses the wish that all National Committees should adopt the text of the IEC recommendation for their national rules in so far as national conditions permit. Any divergence between the IEC recommendation and the corresponding national rules should, as far as possible, be clearly indicated in the latter'.

Regarding the second point, standards for switchgear were developed long before the advent of vacuum switching equipment, they are rooted in other technologies. Even so, it is surprising that the word 'vacuum' does not appear in the table of contents of 56 IEC 1987, and the only reference in the ANSI C37 standard appears in C37.90-1989, 'Alternating-current high-voltage power vacuum interrupters - safety sequirements for X–radiation limits'. It is true that much of what the standards treat is applicable to switchgear generally, regardless of the switching technology involved. Nevertheless, the opinion (which the writer shares) has been expressed that vacuum devices have been forced to conform with rules that have no relevance to the vacuum technology. An example is the temperature limit for contacts. After much (appropriately heated) discussion, this constraint has been relaxed.

10.3 Basic duties and basic concerns

A switching device can be in the closed or the open position. In practice it is likely to spend much more time closed than open. In the closed position the major concern is with the current it is capable of carrying which must be comfortably adequate for the application. The potential problems here are thermal, an overloaded switch will overheat. Care should be taken that the free circulation of air provided by the manufacturer's design is not obstructed. There are frequently different ratings for vented and unvented applications.

Most pieces of switchgear have a temporary overcurrent rating, identified by the manufacturer. If this is likely to be exceeded, a device of higher current rating should be selected.

Switching equipment sees much higher currents under fault conditions, which they may or may not be required to interrupt. In any event, they must sustain these currents until they are removed. For this purpose every equipment has a 'momentary current' rating. The concern in this instance is for the mechanical forces which arise as a consequence of electromagnetic interactions. A typical short-circuit current has an exponentially decaying DC component super-imposed on the AC component as indicated in Figure 10.1. The degree of

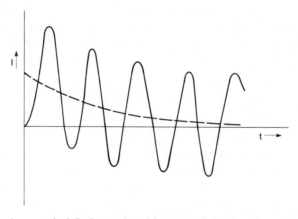

Figure 10.1   Asymmetrical fault current with exponentially decaying DC component

asymmetry depends on the instant in the voltage cycle when the fault occurs [2]. Under the most unfavourable conditions, the peak fault current can approach twice the symmetrical fault current peak. Recognising that electromagnetic forces increase with the *square* of the current, it is apparent that the peak forces for a fully asymmetrical current will be four times as great as for a symmetrical current. These facts apply to all switchgear. Of particular significance to vacuum equipment is the potential for contact popping, a phenomenon discussed in Section 6.2.

Still another consideration is the 'close and latch' rating. The momentary current just described occurs when the circuit-breaker is in the closed position. The breaker itself will initiate the current if it closes into a fault. On such occasions it must not fail to close, hence the close and latch rating. It is described in ANSI/IEEE Standard C37.04-1979 (reaffirmed 1988) 'Standard Rating Structure for AC High-voltage Circuit Breakers Rated on a Symmetrical Basis'.

When a switch is in the open position the concern is with its dielectric integrity; it must not be applied at a voltage higher than its rating. It is not normal practice to leave a circuit breaker in the open position with voltage applied for any appreciable length of time. It is usual to isolate the breaker with disconnecting switches of some kind. With metal clad equipment this isolation is provided by the primary disconnects, the silver-plated, spring-loaded copper fingers that embrace the stabs on the buswork behind the breaker. After it has been opened, the breaker is racked out of position leaving two clear breaks at the primary disconnects of each pole.

In breakers and switches other than vacuum, a strong point is made that insulation co-ordination should be such that external flashover to ground must always occur before internal sparkover between the contacts. The reason for this is that if a breaker or switch in the open position sparks over internally (due to a lightning surge, for example), there is no means available to quench the ensuing arc; the breaker would be destroyed. An external flashover, though inconvenient, can be cleared by the backup breaker, before serious damage has been done. This co-ordination is unnecessary, indeed undesirable, in vacuum. The arc produced by an internal sparkover between the contacts is interrupted at the first power-frequency zero, if not before, since conditions are no different from a normal switching operation. There is a high probability that the event would pass unobserved. An external arc, on the other hand, would put a fault on the system that would have to be cleared by another breaker (as with an oil or gas blast breaker).

The danger of X–rays with an open switch is negligible at normal system voltage, but precautions must be taken when performing hipot tests. More will be found on this subject in Chapter 11.

Important as the closed and open positions may be, the highest stresses are usually experienced during transition between the two, i.e. during closing and opening. It is normal for a switching device to prestrike on closing, that is to say the current is established by a spark/arc, before the contacts physically touch. In some operations - closing into a fault (already mentioned), connecting one capacitor bank to another - the prestrike current may be very high. The consequences of this must be considered in any application.

Opening operations and the interruption of current subject the switching device to thermal and mechanical shock as well as electrical stress following current interruption when the TRV develops. These factors have been discussed in Chapter 6. They will be reviewed again for the various applications about to be considered.

10.4   Load switching

We are considering here the connection and disconnection of loads with relatively high power factor. Such operations present little challenge to vacuum equipment. Figure 4.2 showed how TRV develops following current interruption; the voltage oscillates about its final steady-state value, the instantaneous system voltage. With a fault the power factor is low and the instantaneous system

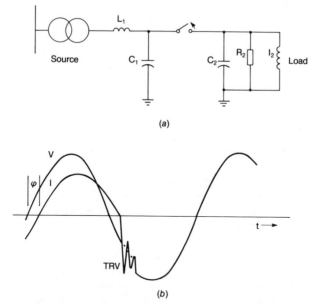

(a)

(b)

Figure 10.2   Interruption of relatively high power factor load

(a) Circuit
(b) Voltage transients evoked

TRV is the difference between V_A and V_B

voltage is near its peak at current zero, so the TRV is high. With a high power factor load, the current and voltage will be almost in phase, thus, when interruption occurs at current zero, the system voltage will be low and the TRV will be low as a consequence. This point is illustrated in Figure 10.2(*b*), which shows the consequences of the load voltage oscillating about zero which is its final value after disconnection.

This transient will experience considerable damping because of the resistive element of the load. The oscillation on the source side V_A, involves the components L_1 and C_1, but that will be very small because the source side voltage will change very little with the shedding of the load. A more detailed discussion of load switching will be found in Reference 3.

The arc voltage in vacuum will be around 20 V during such a switching operation, the arc current will be perhaps a few hundred amperes, and the duration of the arc will be less than half a cycle. Consequently contact erosion will be very small indeed. We can therefore expect a switch to be capable of many thousands of such operations in its lifetime.

10.5 Interrupting fault currents

10.5.1   Selecting a short-circuit current rating

To select the short circuit current rating of a breaker it is necessary to set up a one-line diagram of the system where the breaker is to be located and carry out a

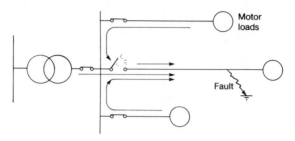

Figure 10.3   Fault current contribution from nearby motors

short circuit study. Generally speaking a terminal fault on the downstream side of the breaker produces the highest short circuit current. If there are motors connected to the bus, as in Figure 10.3, their backfeed contribution to the fault can be important and must be taken into account.

Current asymmetry of the kind shown in Figure 10.1 is to be expected. In the case of a three-phase fault, at least one phase will experience a significant degree of asymmetry. The requirements for breakers to handle asymmetrical currents are spelled out in the standards, ANSI/IEEE Standard C37.04-2979 (reaffirmed 1988) and IEC Publication 56, 4th Edition, address this matter. Asymmetrical currents present no problems to vacuum circuit breakers, provided they are within the breaker rating. Interruption almost invariably occur at the first current zero, even after a fully offset, major loop of current.

Attention should be paid to future development of the power system when selecting a breaker rating. New system interconnections, the extension of a bus and the addition of another transformer, or the local connection of a generating plant furnished by an independent power producer can all raise the short circuit level at a given location. One does not wish to replace the breakers after a few years, with breakers of higher capacity, because the short circuit current level has risen.

10.5.2   The evolving fault

There are occasions when the nature of a fault changes during the time the breaker is opening to clear it. This could occur when a wayward arc causes what was a single line-to-ground fault to develop into a double line-to-ground fault. A multiple stroke lightning flash can also precipitate these so-called 'evolving faults'. The effect is to cause a sudden increase of current during the process of interruption. Such events can destroy some circuit breakers, but not vacuum breakers, which take them in their stride.

10.5.3   Voltage considerations

It goes without saying that the steady-state voltage rating of a switching device should match the system operating voltage. The transient voltages associated with short-circuit current interruption must also be within what the breaker can sustain. This will be assured in most instances if the breaker complies with the appropriate standard.

The TRV that a breaker sees when interrupting a short circuit current depends on the system grounding, the nature of the fault, the asymmetry of the fault current and, most importantly, on the characteristics of the circuit. If the power system and load are effectively grounded, the individual poles of the breaker, though tripped in unison, operate more or less independently. For example, if there is a line-to-ground fault on phase A, pole A interrupts the current against the system line-to-neutral voltage. The power-frequency voltage in Figure 4.2 is this voltage; the maximum TRV could approach twice this power frequency peak. The other two poles would interrupt in sequence, as their current zeros occurred, whatever load current was flowing. A line-to-line fault would be interrupted by the poles on the faulted phases against the line-to-line voltage. This is $\sqrt{3}$ times greater than the line-to-ground voltage, but the duty on the individual interrupters would be reduced because there are two in series.

The situation is somewhat different when the load neutral is ungrounded, as will be apparent from Figure 10.4. As long as there is no fault, the point N will be at or close to ground potential because of the symmetry of the circuit. It is as if N were held there by three equal springs.

For the same reason, the same situation will prevail if a three-phase fault

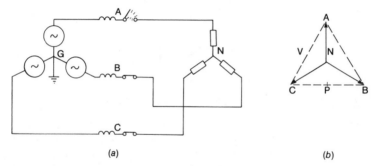

(a)   (b)

Figure 10.4   Interrupting fault on system with ungrounded neutral

   (a) Circuit diagram
   (b) Phasor diagram

occurs and the three poles of the breaker are still closed or arcing. However, once a switching operation is initiated, conditions quickly change. Because the currents in the three phases pass through zero in sequence, one phase, say A, will interrupt first. When this occurs the constraint on N, just described, will no longer exist. Instead N will tend to move to the midpotential P of the two phases to which the load is still connected. If the other poles of the circuit breaker would remain closed, the voltage AP appearing across pole A would be $\sqrt{3}V/2$, so that during the transient period of the TRV it can reach twice this value with a highly inductive load. Under normal circumstances, of course, the other two phases would interrupt shortly after the first (90° later to be precise). Their duty would be relatively less severe, inasmuch as the two remaining phases are in series, with two poles to interrupt against a voltage $\sqrt{3}V$. With a grounded system, therefore, the three poles of the breaker have identical jobs to do, but with an ungrounded system the duty of the first phase to clear is more severe than that of its buddies and more severe than the case of a grounded system. Of course, switching operations occur quite randomly, so we cannot designate which will be the first pole to clear; all must have the necessary capability.

Nature is kind with respect to the TRV experienced by a breaker after interrupting an asymmetrical fault current. As always, current interruption occurs at a current zero and the TRV oscillates about the instantaneous system voltage. Unlike the symmetrical current case depicted in Figure 4.2, the system voltage is not near its peak, thus the TRV is lower in magnitude, as will be observed in Figure 10.5.

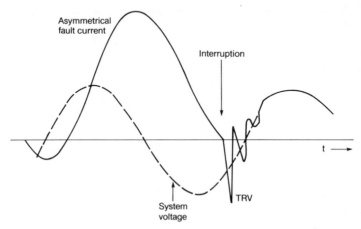

Figure 10.5   Reduced TRV following the interruption of an asymmetrical current

The severity of the TRV depends on its magnitude and rate of rise. This suggests that a symmetrical fault current, followed by a high-frequency TRV would be the most onerous condition for a circuit breaker. Of note in this category is the fault on the secondary side of a transformer, shown in Figure 10.6, a situation that has been studied by Harner *et al.* [4].

The factor which contributes to the high frequency in this circuit is the relatively low value of the transformer capacitance in Figure 10(*b*). The presence of other lines on the secondary side will relieve this situation.

Another condition which can give a high rate of rise of TRV is the short line fault [1], which is a consequence of interrupting a current to a fault a relatively short distance - one kilometre or thereabouts - from the breaker. Superposition is a good way to analyse the TRV experience by an interrupting breaker. Imagine injecting at the switch contacts, at the time of interruption, a current equal and opposite to that being interrupted. The total current is then zero (as it is after interruption) and, more importantly, the response of the system to the injected current, as calculated at the contacts, gives the breaker's TRV. Since the period of interest is short, it is usually sufficient to represent the injected current by a ramp with the appropriate slope (dI/dt). Applying this method to the short line fault situation, one finds a TRV of the shape shown in Figure 10.7. The saw tooth is contributed by the line side of the breaker. The rapid rate of rise of voltage, applied while the arc products are still dispersing, has been a source of trouble for some circuit breakers.

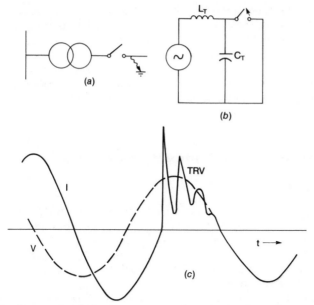

Figure 10.6   Interrupting fault current on secondary circuit of transformer

    (*a*) *Circuit*
    (*b*) *Method of analysis*
    (*c*) *TRV*

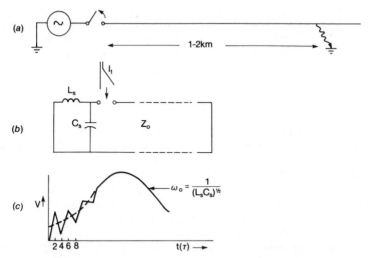

Figure 10.7   TRV experienced by breaker after interrupting short line fault

    (*a*) *Circuit*
    (*b*) *Method of analysis*
    (*c*) *TRV*

Travel time breaker to fault $= \tau$

Short line faults have been mainly studied in connection with overhead transmission lines, however, they can be visualised on distribution lines and so vacuum breakers could be subjected to them. The extremely fast recovery of dielectric strength of a vacuum contact gap after arcing, discussed towards the end of Section 4.1 and quantified by the coefficient ξ in Equation 4.1.1, assures that vacuum switching devices can take short line faults in their stride. The same is true of fault clearing on the secondary side of a transformer, the subject of the immediately prior discussion.

10.6 Interrupting small inductive currents

10.6.1   Concerns

This is by no means a subject of concern for just vacuum switching devices, it ha been a preferential subject in CIGRE on several occasions and the focus of an exhaustive series of CIGRE reports [5], embracing all switching technologies. The operations of interest are the interruption of transformer magnetising current, the connection and disconnection of reactors, and the disconnection of reactor loaded transformers.

The potential problems are the consequences of current chopping and of reignition following interruption.

10.6.2   Current chopping

This phenomenon is discussed in some detail in Section 4.5, where it is shown that the premature interruption of current before current zero (the chop) can lead to an overvoltage. The magnetic energy stored in the inductive elements of the power circuit, when released into the system, can build up considerable overvoltage as it transfers to capacitive elements in the circuit. By equating the stored magnetic energy to the stored electric energy, it is shown (Equation 4.5.1) that, neglecting damping, a voltage

$$I_0 Z_0 \sin \omega_0 t$$

can be generated, where I_0 is the current chopped, Z_0 is the surge impedance of the component in which the current is flowing, and ω_0 is the angular natural frequency of the circuit. It is evident that this voltage will increase directly with the chopping level of the switching device and the surge impedance of the object being disconnected. It was pointed out that this voltage is independent of the system voltage. These facts lead to the conclusion that the most vulnerable pieces of equipment are those with a high surge impedance and low rated voltage, being switched by a device with a high chopping level.

It is my long experience that oil-insulated transformers rated 13.8 kV and higher, that is to say transformers of 95 V BIL or higher, are most unlikely to be damaged when being switched unloaded with a vacuum device having conventional copper–bismuth (Cu/Bi) or copper–chromium (Cu/Cr) contacts. Section 4.2.3 showed that one cannot assign a single value to the chopping level, it varies statistically about some most likely values. But even the occasional high chop is unlikely to result in a damaging overvoltage for the conditions cited. An important contributing factor is the considerable damping that the transformer core losses introduce. The transient produced by a chop is likely to be aperiodic or almost so, indicating that the peak is probably no more than 40 or 50% of what one would calculate ignoring damping. Moreover, Lee [6] has shown that the voltage distribution within a transformer winding due to a current chop is relatively uniform unlike its response to a fast-rising voltage surge where the stress tends to concentrate across the line end turns. As the kVA rating of a transformer is reduced, the capacitance tends to diminish as the magnetising inductance increases. Both these trends tend to increase the surge impedance, but the magnetising current itself is reduced, so that even if the switch chops the current at its peak, I_o is likely to be insufficient to create a serious problem. What is often observed when disconnecting a low kVA rated transformer is that the arc is extinguished very soon after the contacts part, but then the short contact gap is unable to support the transient voltage that ensues, and breaks down. The breaker succeeds in interrupting the reignition current and the TRV begins to rise again. This sequence can repeat several times as indicated in Figure 10.8 which is from a paper by Greenwood *et al.* [7]. Note how the breakdown voltage increases with increasing contact gap. Energy is dissipated with each reignition so that the

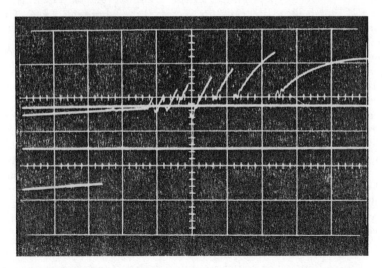

Figure 10.8   Voltage limiting by reignitions following current chop [7]

(*By permission of IEEE*)

highest voltage attained is not nearly as high as the initial prospective peak.

The magnetising current can be much higher on energisation. One might suppose that because there is an opportunity to chop more current, a higher transient voltage might be observed. My somewhat limited experience of this does not bear this out. In any event, it is unusual, in practice, to perform such an operation and caution suggests it would be wise to avoid doing so.

Dry-type transformers of the low voltage rating are more vulnerable to damage from chopping transients. The absence of a liquid dielectric may well decrease the transformer capacitance and therefore increase its surge impedance. Ttransient overvoltages due to current chopping bear no direct relation to system voltage, thus the low voltage rating is an important consideration. Moreover, the BIL and SIL with respect to rated voltage is often lower for dry-type transformers. In my view, you get what you pay for.

There are often mitigating circumstances which reduce the hazard from current chopping. For example, transformers are frequently separated from their switchgear by a length of cable. It does not need much cable to greatly increase the capacitance of the circuit being switched. The capacitance of the cable is roughly 80 pF/ft or 250 pF/m. Thus, 50 ft of cable would add 4nF to the cable transformer capacitance which might well be around this value itself. In that event the surge impedance which might be reduced by a factor of $\sqrt{2}$.

If an application presents some concern, there are a number of options available to ease the situation. One can apply surge capacitors or surge arresters. In either event they should be connected as close as possible to the terminals of the transformer being protected. A small amount of resistance in series with the surge capacitor is advisable. This assists in damping transient oscillations. It is extremely important that the resistor used be noninductive and that the leads be as short as possible. Capacitance *directly across the vacuum switch* is not a particularly good idea; it tends to increase the chopping level. Critical transformers, especially if they are switched frequently, should be protected by a surge arrester as a matter of prudence.

Another option is to use a switch or breaker whose contacts have a low chopping level; the strong dependence on contact material is discussed in Section 4.5.2. It is now claimed [8] that a lower chopping level can be obtained without sacrificing other power switching capabilities.

The switching of shunt reactors is different from unloaded transformers in that the current is significantly higher. This provides an opportunity for the breaker to chop a higher current if it is capable of so doing. The fact that the reactor current is higher means that its inductance, and therefore its surge impedance, is lower. The current ratio of comparably rated reactors and transformers is roughly the load current to the magnetising current of the transformer. This may be 1000:1, thus with comparable capacitances, the surge impedances would be roughly in the ration 1:30. It would therefore take a much greater chop to produce a given surge in the reactor.

10.6.3   Reignitions when switching unloaded transformers and reactors

The consequences of reignitions when interrupting transformer magnetising current have been discussed in connection with Figure 10.8. It was noted that such action can be beneficial in that it limits the chopping voltage transient, and that the relatively low-frequency transient produced by a current chop distributes more or less uniformly along the transformer winding. However, when the voltage collapses at a reignition, it is as if a very steep-fronted surge of opposite polarity, and magnitude equal to the reignition voltage, were suddenly impressed on the transformer. This can produce a very nonuniform voltage distribution within the transformer winding, causing high stress on the interturn insulation of the line end turns [9]. It is rather like applying a chopped wave test to the transformer [10]. In the present context, the magnitude of these assaults is unlikely to be dangerous except where dry-type transformers with reduced dielectric strength are concerned.

Reignitions during the switching of reactors can be a horse of a different road clearance. A sequence is described in Section 4.6.1 and illustrated in Figures 4.30 and 4.31, wherein reignitions lead to currents with multiple frequencies. The combined current (Figure 4.31) has current zeros, at which the switching device may well interrupt. The energy that can be trapped in the inductive elements of the circuit by such a high frequency arc extinction can be considerable. It can be much greater than would be the case with normal current chopping. The transient voltage produced is correspondingly greater.

The sequence of reignition and clearing can become repetitive, and since the breakdown voltage of the contact gap increases as the contacts move apart, the reignitions tend to be from progressively higher voltages. The phenomenon of voltage escalation by repetitive reignitions and clearings has been described by Greenwood *et al.* [7], as it affects motor switching, and by Itoh *et al.* [11]. More recently Greenwood and Glinkowski [12] have analysed the events in some detail, Figure 10.9 is taken from their paper. They identified two conclusions for such a sequence: type A in which the breaker clears and type B in which the current carries over into another power frequency loop; Figure 10.9 is an example of the first. Because the cause of the overvoltage is the same as that for current chopping, namely, the trapping of magnetic energy in inductive elements of the circuit and its subsequent release to capacitive elements, the author coined the phrase 'virtual current chopping' [13] to describe it, as noted in Section 4.6.1.

It is inevitable that reignition will occur from time-to-time when switching reactors because there will be occasions when the contacts of a pole separate just before a current zero but the current is interrupted, nevertheless. The reignition occurs because the contact gap is very short when the TRV is applied, too short in fact for the gap to support the transient voltage. There is therefore a chance of multiple reignitions and voltage escalation. Some manufacturers recommend the application of surge suppressors because of this, other manufacturers do not. The wise course is to consult with the manufacturer and follow his recommendations. If in doubt, install suppressors.

A reignition in one pole of the switch can cause current to be interrupted in one or both of the other poles if they are arcing at the time, since the high-frequency component of the reignition current returns through the other poles. One company has sought to avoid this problem by 'advanced opening' of one pole. By design, one pole of the switching device opens 8–10 ms ahead of the other two. If it reignites in the course of opening, it will not cause current interruption in the remaining poles, since they are closed. A reignition in the advanced pole after the other two have opened is quite unlikely, because the first pole will have achieved full contact gap by that time.

10.7 Capacitance switching

10.7.1 Energising a capacitor bank

The concern here is for the magnitude of the inrush current when the switch closes. The basic elements of the circuit are shown in Figure 10.10(*a*), they are the inductance of the source L_s, and the capacitance of the bank C. Together

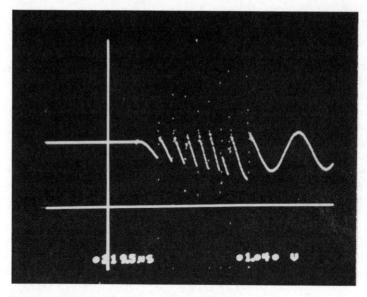

Figure 10.9(a)   Typical reignition sequence with voltage escalation
Length of graticule lines: horizontal = 1ms, vertical = 260 kV

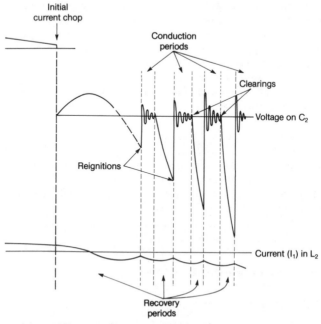

Figure 10.9(b)   Detailed portrayal of sequence of reignitions and clearings [12]
(By permission of IEEE)

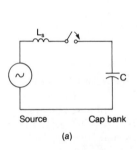
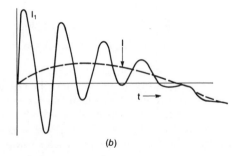

Source   Cap bank

(a)   (b)

Figure 10.10   *Inrush current on energising a capacitor bank*

    (a) Significant elements of circuit
    (b) Current waveform

they form a series resonant circuit. Thus, on closing, there is natural frequency current given by

$$I_1 = \frac{V_1}{Z_{01}} \sin \omega_{01} t \qquad (10.6.1)$$

where Z_{01} is the surge impedance of the circuit, $= (L_s/C)^{1/2}$, ω_{01} is the angular natural frequency $= (L_sC)^{-1/2}$, and V_1 is the voltage across the switch at the moment of energisation. This current is usually much greater than the power-frequency current on which it is superposed. An example will make this clear. Consider a capacitor bank rated 6 MVAR at 13.8 kV. Its rated current is

$$\frac{6 \times 10^6}{\sqrt{3} \times 13,800} = 251 \, \text{A rms}$$

This means that

$$X_c = \frac{13,800}{\sqrt{3} \times 251} = 31.74 \, \Omega$$

Assuming 60 Hz

$$C = \frac{10^6}{377 \times 31.74} = 83.6 \, \mu\text{F}$$

Assume $L_s = 1.057$ mH which corresponds to a symmetrical short circuit current of 20 kA at 13.8 kV. The surge impedance is then

$$Z_{01} = \left[\frac{1.057 \times 10^{-3}}{83.6 \times 10^{-6}} \right]^{1/2} = 3.556 \, \Omega$$

The natural frequency is $\omega_{01} = (1.057 \times 10^{-3} \times 83.6 \times 10^{-6})^{-1/2} = 3364$ rad/s or 535 Hz. If the bank is discharged at the time of energisation, and if closing occurs at the peak of the voltage cycle, the peak high frequency current from eqn. 10.6.1

will be

$$\frac{V_1}{Z_{01}} = \frac{\sqrt{2} \times 13,800}{\sqrt{3} \times 3.556} = 3169 \, \text{A}$$

which is considerably more than the 60 Hz current. The current is illustrated in Figure 10.10(b). As mentioned earlier, the circuit is likely to be completed by a prestrike before the contacts touch mechanically. If this occurs at a relatively long contact gap, it is possible that a high-frequency current zero may occur during the prestrike. Should this happen, it is entirely possible that the switch will interrupt the current momentarily because of its very fast rate of recovery (the factor ξ in Section 4.1, Equation 4.1.1).

Sometimes a capacitor bank is switched out and then switched back in again a short time later, before the charge trapped on the bank has had time to discharge significantly. It may happen that the prestrike occurs when the bank and source have opposite polarities. Thus, V_1 in Equation 10.6.1 can approach 2×line-to-neutral peak system voltage, in which case the high frequency current would be double that calculated.

The events just described are interesting but present no real challenge to the vacuum switching device. Pickup from the inrush current might conceivably cause damage to neighbouring control circuits if they are not well screened.

10.7.2   Bank-to-bank capacitor switching

Capacitor banks are often switched onto a power system as the load builds during the day to maintain the system voltage which would otherwise be depressed by the regulation of the source. The reverse procedure is followed when the load declines. A typical arrangement, showing three sections of capacitor, is illustrated in Figure 10.11.

The capacitor bus is protected by a circuit breaker which takes care of faults on that section. The individual operations of connecting and disconnecting capacitors is performed by load break switches.

Suppose the situation of Figure 10.11 exists, the first capacitor bank is energised and we are about to energise the second. Once again there will be an inrush current plus a power frequency current from the source, but in addition, capacitor 1 will dump charge into capacitor 2. The banks are likely to be physically close together, so unless some deliberate steps have been taken to make things otherwise, there will be little impedance between them. The local surge impedance will therefore be quite low and the corresponding current when capacitor 1 shares its charge with capacitor 2 will be quite high.

Assume banks 1 and 2 are both 6MVAR and that the loop inductance L_2, of their connecting bus is $40\mu\text{H}$, then the inrush current from bank-to-bank is

$$I_2 = \frac{V_2}{Z_{02}} \sin \omega_{02} t \tag{10.6.2}$$

where Z_{02} is the surge impedance of the loop involving capacitors 1 and 2 and L_2

$$Z_{02} = \left[\frac{2L_2}{C}\right]^{1/2} = \left[\frac{80}{83.6}\right]^{1/2} = 0.97 \, \Omega$$

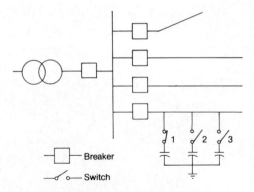

Figure 10.11   Substation with sectionalised capacitor bank

ω_{02} is the angular natural frequency of this circuit

$$\omega_{02} = \left[L_2 \frac{C}{2}\right]^{1/2} = [40 \times 21.8]^{-1/2} \times 10^6 = 3.386 \times 10^4 \,\text{rad/s or } 5.39 \,\text{kHz}$$

If, as before, C_1 is at peak voltage and C_2 is discharged, the peak value of I_2 is

$$I_{2_{\text{peak}}} = \frac{13.800\sqrt{2}}{\sqrt{3} \times 0.978} = 11,520 \,\text{A}$$

As before, if C_2 has the opposite polarity of charge, this current peak can approach 23 kA! Clearly, this is something to be reckoned with as far as the switch is concerned.

There is such a wide separation between the natural frequency of this charge-sharing event and the oscillation between the capacitors and the source, that the first is more or less over before the second has started. In the example cited, where the capacitors are equal, conservation of charge assures that if their initial voltages are $V_1(0)$ and $V_2(0)$, they will come to a common voltage

$$V(\infty) = \frac{V_1(0) + V_2(0)}{2} \tag{10.6.3}$$

having due regard to the signs of $V_1(0)$ and $V_2(0)$. This common voltage is different from the instantaneous source voltage, thus a second surge of current from the supply causes the capacitor bus voltage to oscillate about the system voltage. The frequency will be $\omega_0 = 2379$ rad/s, or 378 Hz, since there are now two banks of capacitors involved. It is worth noting that the presence of the connected parallel feeders will help to damp this oscillation.

An oscillogram taken on the occasion of an event somewhat similar to that just described is shown in Figure 10.12. On this occasion there were quite a number of *cables* connected to the bus when an uncharged capacitor bank was energised. The capacitor voltage cannot suddenly change, so the bus voltages momentarily drop to zero when each pole of the breaker closes (they do not close simultaneously). Thereafter, the cables equilibrate their voltage with the capacitor bank; this is the high-frequency oscillation observable immediately

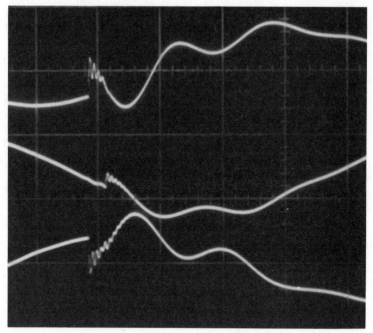

Figure 10.12   Voltage transients on substation bus to which many cables are connected, occasioned by switching onto bus of uncharged capacitor bank

1 div = 20 kV = 2ms

after closing. Subsequently, the cables and the capacitor bank are restored to the supply voltage by the quite evident, much lower frequency oscillation which follows.

What lessons are to be learned from this discussion? High magnitude, high-frequency inrush currents are to be expected when energising capacitor banks, especially when switching one bank against another. The switch will almost certainly prestrike, so that the high current will pass in the arc before the contacts physically mate. This can cause considerable contact erosion and shorten the life of the device. There is some evidence that this treatment also degrades the dielectric strength of an interrupter [14].

High currents mean high mechanical forces on the interrupter and its connections. In this case, these are short impulsive forces. The matter of electromagnetic pickup in nearby control circuits becomes increasingly important as the magnitude of the current involved increases [15].

Inrush currents can be mistaken for fault currents and cause misoperation of substation relays if the relays are inadequately co-ordinated. Finally, capacitor banks are constructed from individual capacitor cans which are connected in parallel (and series too, depending on the voltage). Individual cans or series strings are fused to avoid tripping off a bank because of the failure of a single unit. It is unlikely, but surely possible, that fuses can be blown by repeated inrush currents.

What can be done to relieve the situation? There are solutions but none is especially attractive. The first is to insert impedance in the circuit to reduce the magnitude of the inrush current. A preinserted resistor will essentially eliminate any oscillation, but if it remains in the circuit it will cause a significant steady-state loss. For example, in the installation cited above, where the surge impedance for bank-to-bank switch is approximately 1 Ω, one might wish to insert a 2 Ω resistor. For a 6MVAR bank at 13.8 kV this would mean a loss of 125 kW/phase! This is clearly unacceptable. Arrangements must therefore be made to short out the resistor when the transient is spent. This requires another pair of contacts which must open and close, and retain their integrity in every way through the life of the device. This is not a trivial complication.

An alternative is to insert a reactor in the circuit. Again in the cited example, it would be necessary to quadruple the circuit inductance to reduce the magnitude of the magnitude of the inrush current by 50% (not a great reduction). This would mean adding $3\times40 = 120\mu H$/phase. This too is not a trivial cost item and would have some losses, though much less than the resistor. The pros and cons of preinserted resistors and reactors has been discussed by O'Leary and Harner [16].

Another possibility is *point of wave switching*, which means selecting a point to close on the power-frequency voltage wave when the voltage across the contacts is low. The consequences of keeping V_1 and V_2 low in Equations 10.2.1 and 10.2.2, respectively, is obvious. This requires independent pole operation for the switching device and a relatively sophisticated sensing package to determine the precise instant to initiate closing. It also requires that the mechanism be consistent in its closing time throughout its life. Moreover, a switching device has a way of making its own decisions, at least to some extent. There is a propensity to prestrike on closing. This is more likely to occur when the intercontact voltage is high than it is when it is low, which is contrary to what we desire. Fast closing is helpful in this regard, but problems such as contact bounce (Section 6.3) set limits on this.

One concludes from this discussion that there is no such thing as a free lunch. At the cost of some flexibility some prudent steps can be taken to avoid the worst conditions. For example, closing can be blocked unless the incoming bank is essentially discharged. It is worth pointing out that if the capacitor bank neutral is ungrounded, the closing of the first pole simply charges the stray capacitance at the neutral and raises the potential of the entire bank. This usually precipitates a prestrike at another pole, thereby connecting two phases in series. This causes an inrush current of course, but the surge impedance of the circuit is higher, involving as it does the series connection of the inductance and capacitance of two phases.

10.7.3 Disconnecting a capacitor bank

The disconnection of a capacitor bank is a simple, straightforward operation provided the switch does not restrike or reignite*. Such an occurrence is a possibility because the contact gap remains cyclically stressed for a considerable period after current interruption. How this comes about in a single phase operation will be evident form Figure 10.13. When the capacitor was connected, the bus voltage rose; this is the so-called *Ferranti rise*, occasioned by the leading

---

* A dielectric breakdown in the first half cycle after interruption is a reignition; thereafter it is a restrike.

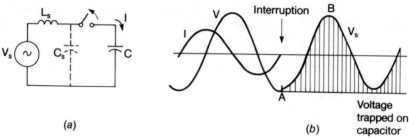

Figure 10.13   Disconnection of single-phase capacitor bank

(a) Circuit
(b) Current and voltage waveforms

current drawn by the capacitor bank flowing through the inductive reactance of the source. When the capacitor is disconnected, there is a corresponding fall in the bus voltage. The transition occurs through a transient oscillation involving the source inductance L_s, and the stay capacitance of the source C_s. This small transient is clearly visible at A in Figure 10.13(b). Because the current is capacitive and leading the voltage by 90°, the voltage is at or close to a peak when the current passes through zero and is interrupted. Peak voltage is thereby trapped on the capacitor and a high voltage will remain there for a considerable time, until charge slowly bleeds off through the internal discharge resistors. The other side of the switch, on the other hand, will follow the power-frequency voltage of the source. The voltage seen by the switch is therefore that shown shaded in Figure 10.13(b), and neglecting the brief transient at A, can be described by

$$V = V_P(1 - \cos\omega t) \qquad (10.6.4)$$

So it is that the switch experiences a maximum voltage of $2V_P$, or close to this value, each cycle after interruption. If the source and capacitor bank have solidly grounded neutrals. V_P is peak line-to-neutral voltage to which we assign 1 pu. If the capacitor bank neutral is isolated, interruption of current in the first pole to clear causes a shift in the neutral voltage which is quite similar to that described in Section 10.6.4 and illustrated in Figure 10.4. As in that situation, so in this, the three poles do not see the same duty. The first pole to clear has a recovery voltage of 2.5 pu the voltages across the other poles are reduced.

The cyclically pulsating voltage stress across the open contacts of a switch, following the disconnection of a capacitor bank, has always been a challenge to switchgear engineers because of the opportunities it provides for a restrike. It is particularly relevant that capacitance switching is a frequent operation, banks are connected and disconnected daily, and sometimes more frequently, so an installation can quickly rack up a large number of operations. What are the implications for vacuum devices? Generally speaking, vacuum switches and breakers have a high dielectric strength and as noted in Section 4.2.2, they achieve their recovery very quickly indeed, significantly quicker that other switching technologies. This is especially true when the arc in diffuse as would be the case with capacitance switching. However, it was also noted in Section 3.1 that breakdown in vacuum tended to be more statistical than other switching technologies, and in Section 3.3 it was observed that breakdown is much dependent on the material of the contacts and the condition of the contact surfaces. As far as contact surfaces are concerned, these are continually changing as switching operations take place. In general, interrupting a current of a few hundred amperes has a salutary effect on dielectric strength, but subjecting the switch to severe inrush currents on closing causes some deterioration. Thus, it has been observed, and reported by Osmocrovic [14], that dielectric recovery is better when switching individual capacitor banks than it is when engaging in bank-to-bank switching. This discussion should not be construed as indicating that switching capacitor banks with vacuum interrupting devices is a risky business. This is not so; modern vacuum interrupters perform very well. It is to say that an occasional restrike may occur and the consequences when it does are now examined.

A restrike is a form of closing operation, therefore the considerations of Section 10.7.1 can be applied. If a restrike occurs at B, for example, in Figure 10.13(b), V_1 would be 2 pu for a neutral grounded bank and 2.5 pu for a bank with isolated neutral. This directly affects the magnitude of the restrike current. What is special to vacuum is its ability to interrupt high-frequency currents as discussed in Section 4.6. Now examine the consequence of such a high-frequency current interruption.

Suppose that a restrike does occur at point B in Figure 10.13(b), reconnecting the charged capacitor bank to the source. Inrush current at frequency $1/2\pi(L_sC)^{1/2}$ will surge into the capacitor, causing its voltage to swing to approximately as far above the instantaneous system voltage as it started below. Figure 10.13(b) is continued from point B, but with an expanded time scale, in Figure 10.14. The attainment of this voltage of 3 pu coincides with a high frequency current zero, at which the switch may clear, trapping the 3 pu voltage on the capacitor. If this should happen, the source side of the switch would oscillate about the instantaneous source voltage (-1 pu) at a frequency determined by L_s and C_s. If the switch survives this without reigniting, the source side will continue following the dictates of the source voltage, so that half a cycle later (and every cycle thereafter) there will be 4 pu across the contacts. If a second restrike should occur at one of these times, point D for example, the capacitor voltage would be swung to -5 pu, as indicated in Figure 10.14.

The sequence portrayed in Figure 10.14 is idealised. For example, in practice restrikes will not always occur precisely at the voltage peak, so that voltage, if it escalates, does so more slowly. What is pointed out here is the potential for generating high-voltage transients which can damage capacitors, or the insulation of other connected equipment, or cause a flashover somewhere.

It is possible to guard against destructive overvoltages by installing surge arresters which clamp the voltage at a predetermined level. However, it is important to carry out a transient analysis of the situation before selecting the surge suppressor. In particular, one must compute the energy to be dissipated. It can be quite high in such applications.

10.7.4 Switching cables

Cables possess a considerable amount of capacitance and behave somewhat like capacitors when they are switched. The fact that they are physically long means

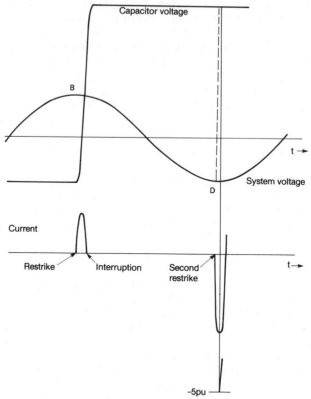

Figure 10.14   Reignitions during disconnection of capacitor bank

that they can support travelling waves which means that their manner of charging and discharging is somewhat different from a capacitor bank. If a cable is energised from a stiff bus when the instantaneous voltage across the switch is V, a square wave of voltage of magnitude V will spread down the line, associated with a current of V/Z_0 where Z_0 is the surge impedance of the cable. The wave will charge the cable capacitance to V with an amount of energy $1/2 \, CV^2$ joules, where C is the total capacitance of the cable. At the same time, the current puts $1/2 \, LI^2$ joules into the cable's inductance during the course of the wave's progress from end to end of the cable, where L is the total inductance of the cable. It is readily shown that $1/2 \, CV^2 = 1/2 \, LI^2$. If the remote end of the cable is open, the magnetic energy as it arrives is converted into electric energy, the current goes to zero and the voltage doubles. Put another way, the reflected voltage wave from the open circuit has the same sign as the incident wave, whereas the reflected current wave has the opposite sign. Various instants during the early stages of energisation are depicted in Figure 10.15; the voltages at the open end of the cable and at the middle of the cable, as a function of time, are shown in Figure 10.16. Time is measured from the instant of closing and is in units of τ, the time for a wave to travel the length of the cable. In practice the waves are attenuated because of losses, they are also distorted, because the

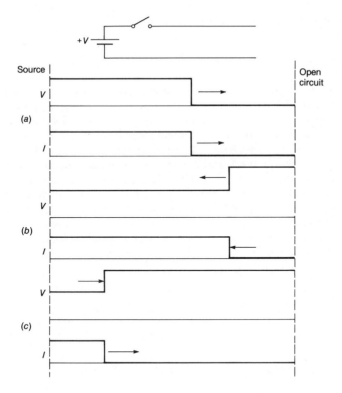

Figure 10.15   Travelling waves on open-circuited cable at several instants after energising

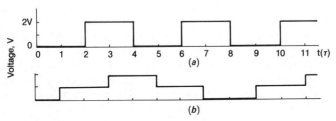

Figure 10.16   Voltages on cable following energisation

   (a) At remote end
   (b) At middle of cable

different component frequencies travel with different velocities. They are also influenced by the impedance of the source. If the source is AC, the magnitude of the initial wavefront is determined by the point in the cycle when current is established. Thereafter, all waves from the source end are such as to reconcile the voltage on the cable with the instantaneous voltage of the source. The wave steps quickly disappear into the power frequency voltage and charging current.

A cable typically has a surge impedance of 30 Ω, thus, if it is energised at the peak of a voltage cycle of 13.8 kV, the initial current is

$$\frac{13,800\sqrt{2}}{\sqrt{3} \times 30} = 376\,\text{A}$$

which presents no problem at all to a vacuum switching device. When the cable is disconnected, the current to be interrupted, if the remote end is an open circuit, is just the charging current of the cable. For a 13.8 kV cable this is of the order of 1A/km, so the current to be interrupted is small. Like a capacitor, it will retain its charge when disconnected. A restrike would set off travelling waves as described under energisation, and current zeros can occur in the restrike current, but it is not the author's experience that voltage escalation occurs.

There are situations, not covered by standards, where higher than normal stresses can occur. One such situation is shown in Figure 10.17; a high voltage cable is being switched from the low-voltage side of a transformer. The condition can arise under certain circumstances, as when a high voltage breaker fails to trip. If breakers 2 and 3 in Figure 10.17 are called upon to open to isolate the

LV network   HV network

Figure 10.17   Switching high-voltage cable from low-voltage side of transformer

cable, and breaker 2 fails to do so, the operation may require that the cable be isolated from the LV network side by opening breaker 1 . The capacitive current will be much higher because of the turns ratio of the transformer, but surely not too great for breaker 1 to handle. However, the high capacitive current, flowing through the leakage reactance of the transformer causes a high-voltage condition on the cable against which breaker 1 must switch. The transformer cannot maintain a high direct voltage at its terminals. It will pull magnetising current until it goes into saturation, at which time the cable voltage will flip its polarity. Indeed, it will continue to flip its polarity, cyclically, applying a square wave recovery voltage to the breaker. The magnitude of this wave will diminish and its period will increase as energy is dissipated in losses. Special situations like this should be discussed with the manufacturer.

10.7.5   Line dropping

The disconnections of unloaded transmission or distribution lines is akin to capacitance switching. However, line conductors have appreciable capacitance to each other as well as to ground. The total capacitance per phase, line-to-line and line-to-ground, is the positive phase sequence capacitance, C_1. That part involving ground only is the zero sequence capacitance C_o. The ratio C_1/C_o is never less than 1. It is infinite when the neutral is isolate and approaches unity asymptotically as C_o increases. The value of the C_1/C_o ratio affects the voltage across the switch after clearing. Specifically, when $C_1/C_o = 1$, corresponding to the solidly grounded case, the maximum voltage across the switch is 2V per unit. The voltage increases with increasing value of C_1/C_o approaching 3V per unit for the completely isolated neutral case. This is discussed on pp. 138 and 139 of Reference 1.

10.8 Recloser applications

Reclosing requires multiple operations within a relatively short time, and many sequences of operations during the life of the device. It is therefore an ideal application for vacuum which has low energy dissipation within the interrupter during any operation and consequently a very long life. The function of a recloser is to protect distribution lines and distribution equipment. A complete recloser comprises the switch itself and its control package, which may be integral with the switch or a free-standing unit connected to the switch by a cable. This self-controlled combination trips open on overcurrent (either phase or ground faults) and then recloses automatically. If the overcurrent is temporary the automatic reclose restores normal service. If the fault is permanent a preset number of trip and reclose operations are performed to lockout.

Opening sequences can be all fast, all delayed, or any combination of fast operations followed by delayed operation. The total is usually limited to four or five. Fast operations clear temporary faults before branch-line fuses can be damaged. Delayed operations allow time for fuses or other down-line protective devices to clear so that permanent faults can be confined to smaller sections of the line.

Standards [18,19] prescribe the recloser's symmetrical interrupting capability at the operating voltage, and also the reclosing capability (the allowable current is reduced as the number of operations increases). Details of this determination are given in References 18 and 19. Meeting these requirements presents no challenge for vacuum reclosers.

10.9 Generator breakers

Generator breakers have a high continuous current (perhaps 25 kA and a high interrupting capability (250–750 kA). They are to be found in nuclear power plants where it is often required to continue supply to the station auxiliaries from the power system while the generator itself is shut down. The function has been served traditionally by massive air blast breakers. Vacuum breakers of these dimensions have not been produced; they may be technically feasible but are probably not economically viable. The high continuous current is not a natural match for vacuum with its butt contacts and heat extraction by way of the contact shanks.

Vacuum is suitable for small generators, the voltage is right and a continuous current of 3 or even 4 kA can be achieved if finned heat exchangers are closely coupled thermally to the contacts. Vacuum breakers are being used in such applications for generators of the order of 100 MW.

There is a trend at the present time (early 1990s) towards machines in the

200–300 MW range rather than the 800–1200 MW of former times. SF_6 generator breakers are available for this class of generator, but they are relatively big devices. This might be an opportunity for two or more vacuum interrupters to operate in parallel. This subject is visited again in Section 10.13.2.

In certain circumstances a generator fault current can flow for several cycles without a current zero. This presents a problem to the vacuum interrupter since its arc voltage is too low to materially change the current waveform. If the application allows for a fast bypass switch to normally operate in parallel with the vacuum interrupt (and carry virtually all of the load current), this switch could commutate the current into the vacuum device which could then interrupt any likely load current or any fault current up to 40 or 60 kA, depending on its design. Putting a resistor in series with the vacuum interrupter could assure an early current zero.

10.10   Arc furnace switching

Because arc furnace switching duty is highly repetitive, it is an ideal candidate for vacuum switchgear. I recall the pre-vacuum days when air-magnetic circuit breakers were used for this function. A furnace installation required at least two breakers, one that was in operation and the other that was being refurbished.

Ore is smelted and metals are alloyed and refined by melting them with the intense heat of an electric arc, using graphite electrodes. Such installations usually operate at low voltage and high current and are consequently fed by a step-down furnace transformer of relatively high impedance. The switchgear is located on the supply side of the transformer where it can be expected to interrupt the transformer magnetising current 50–100 times per day. Switching is required when the transformer is de-energised for tap changing, when taking melt samples, or when adding alloys. In addition, cave-ins will occur, when the material in the furnace effectively shorts out the electrodes. I prefer to call this a long circuit rather than a short circuit, because the impedance of the transformer and the leads to the furnace limit the current, though it is still substantial.

Arc furnace installations are characterised by low power factor, therefore shunt capacitors are normally used for compensation. Sometimes the capacitors are switched with the furnace, but mostly they are not. In some instances static VAr compensation (SVC) is used, especially where the bus is rather weak. One may expect to see more of this as utilities become more sensitised to power quality and cannot afford the disturbances associated with simple capacitor schemes. The SVC capacitors are all tuned to be low harmonic filters, thereby relieving them of some of the higher frequency transient stress. One must be alert to possible problems with the harmonic filter reactors. Repeated energisation of the furnace transformer subjects the filters, especially the second harmonic, to severe electromechanical stress from the highly distorted inrush current.

The circuit breaker has no problem of itself under these conditions, but conditions are conducive to high frequency current interruption with the attendant possibilities of energy trapping and the generation of overvoltages in accordance with the sequence described near the end of Section 10.5. Prudence therefore suggests that the transformer be protected by surge arresters. An iron-clad way of doing this is illustrated in Figure 10.18.

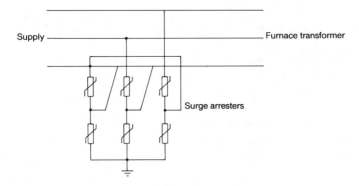

Figure 10.18   Secure way of protecting an arc furnace/transformer installation against voltage surges

10.11 Motor switching

10.11.1   Introduction

Vacuum switchgear is a popular choice for switching motors (vacuum contactors for smaller motors and vacuum circuit breakers for the larger sizes). In many instances the switching operations are frequent. Some large motors are switched directly on line, others require a reduced voltage start. Historically, the strength of motor insulation has not been as great as that of transformers because motors are not exposed to as much abuse from transient overvoltages, especially lightning, as are transformers. The advent of vacuum and the dielectric failure of a number of motors caused a flurry of investigative activity in the 1980s [20–24].

10.11.2   Energising a motor

When a switch closes, the voltage across it disappears, only to reappear instantaneously across the circuit elements on either side of it. The voltage divides in proportion to the surge impedances on either side. A motor is frequently connected to its switchgear by a cable, so the voltage impressed on the load side when the switch closes, travels down the cable to the motor as a wave. This is illustrated in Figure 10.19. If the source impedance is low (if there are many other cables connected to the bus), a high fraction of the voltage V, initially cross the closing switch, travels as a wave to the motor.

The surge impedance of the motor is high compared with that of its cable, so the reflected wave has the same sign as the incident wave. In the limit, if the motor is treated as an open circuit, the wave voltage will double when it reaches the motor terminals. Moreover, in the simple approach followed, it would have a steep front, a square wave in the limit. Analysis shows [20,21] that the impact of such a wave on a motor winding puts severe stress on the first few turns of the motor winding, and notably on the interturn insulation which is not as substantial as the ground wall insulation.

In practice there are factors which tend to reduce the severity of the voltage

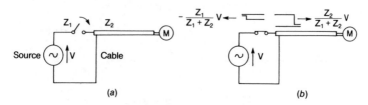

Figure 10.19   Travelling wave initiated in a cable to a motor following closing of its switch

   (a) Before closing
   (b) After closing

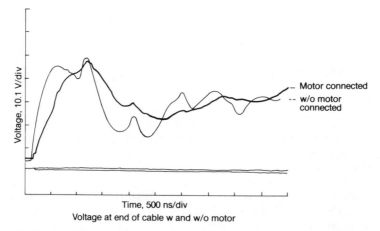

Voltage at end of cable w and w/o motor

Figure 10.20   Modification of the voltage at the open end of a cable when it is being energised, brought about by entrance capacitance of motor [24]

transient: the small inductance of the switchgear buswork reduces the steepness of the front [25]; still further reduction is wrought by the entrance capacitance of the motor itself which is apparent from Figure 10.20.

   As a consequence of such factors it has been established that there is little cause for concern for the effects of normal energisation transients on the insulation of modern motors [24]. Conditions could be more severe if a motor is being transferred from one bus to another. After being disconnected from the first bus, its voltages will tend to slip back as the machine decelerates. On reconnecting to the second bus, the phase of the motor EMF could be vastly different from that of the new bus - they could be 180° out of phase - causing the stress just described to be correspondingly increased.

10.11.3   Disconnecting a running motor

A running motor continues to generate an EMF after it is switched off since its flux takes some time to decay and of course, its speed changes only slowly as it decelerates. Thus, the voltage across the switching device builds up very slowly (apart from a small transient due to the supply regulation) as the applied voltage of the system and the back EMF of the motor drift apart. Current interruption is therefore trivial.

10.11.4 Aborting a motor start

Tripping a motor just after it has been energised and before it has gathered speed is not an advisable procedure. The machine has no back EMF and therefore behaves somewhat like a reactor. Moreover, it is possible for a reignition to occur if the contacts separate close to a current zero. Indeed, there could be a sequence of reignitions with voltage escalation [12] as described in Section 10.6.3. A reignition is a form of closing operation, so the comments of Section 10.11.2 apply. However, there is potential for the voltage V in Figure 10.19 to be considerably higher than the power frequency peak voltage, causing corresponding increases in the stress on the motor insulation.

10.11.5 Recommendation

Because of the possible exposure of motors to potentially dangerous switching surges, it is desirable to protect bigger machines and certainly critical ones. The cost of the protection is small compared with the cost of repairing or replacing a damaged motor. A combination of surge capacitor and surge arrester is ideal, the first being used to slope the front of the surges, the second to limit the surge magnitude. Harder [26] gives some guidance on the selecting of arresters. The importance of closely connecting the protective device with the motor being protected is stressed by McLaren and Abdel-Rahman [27].

10.12 On-load tap changing

Once again the long life and low maintenance characteristics of vacuum switches make them well suited for the frequent switching operations of transformer and phase angle regulator on-load tap changers. The interrupters function perfectly well under oil, and it is the author's experience that they do not present a hazard even if they fail. There must be unrestricted movement of the oil when the bellows move.

10.13 Unusual applications

10.13.1 HVDC circuit breakers

There has not as yet been a great call for HVDC circuit breakers, but the advent of meshed systems will make their use more likely. Prototypes for this purpose have been built [28, 29], a photograph of one appears in Figure 10.21. The principle of its operation is described in Section 4.7 where a diagram of its circuit will be found in Figure 4.36; a more comprehensive description is given in Reference 30.

The direct current is forced-commutated by a precharged capacitor and the entrained inductive energy is absorbed in surge suppressors. Such an

Figure 10.21   Experimental HVDC vacuum circuit breaker (1968) designed to interrupt 5 kA at 100 kV with the author alongside

(Courtesy of General Electric Co.)

arrangement would be ideal for a metallic return transfer breaker [31] used for commutating current out of the earth when a bipolar line is to be operated unipolar with both conductors, as might be the mode if a converter is out of service.

The vacuum approach has not so far found favour with other HVDC applications because of the multiple series interrupters required. The job can be done more readily by airblast with far fewer units [32]. However, this situation could change as airblast breakers become less available and as the current to be interrupted increases. The airblast approach depends on creating an instability in the circuit breaker arc and it is by no means certain that this can be accomplished at currents above a few thousand amperes. The forced-commutation approach is a positive, calculable action. There is no obvious limit to the current that can be handled.

10.13.2   *Parallel operation of vacuum interrupters*

Parallel operation of traditional interrupters is impossible because the negative V/I characteristic of gaseous arcs makes their parallel operation inherently unstable. If the current in one branch increases, the arc voltage falls, causing the automatic transfer of still more current from other parallel paths, so that one path quickly acquires all the current. The arc voltage of a vacuum arc changes little with voltage over a wide range (Section 2.3) and such change as occurs is for arc voltage to increase as the current increases. The author made extensive tests in 1959 to verify this action. Figure 10.22 shows the result of one such test.

It is very important that the contacts of the parallel interrupters open simultaneously or virtually so. The first to open develops an arc voltage V_a, which proceeds to commutate current out of its path and into that of its neighbours at a rate

$$\frac{dI}{dt} = \frac{V_a}{L} \qquad (10.12.1)$$

where L is the parallel inductance of the other paths. Thus if $V_a = 20$ V and $L = 1$ μH, $dI/dt = 20$ A/μs. A modest artificial increase in L by inserting small reactors between the parallel paths, can greatly reduce this rate of current transfer; 56 μH was used in Figure 10.22.

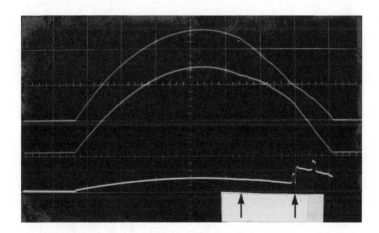

Figure 10.22   Early experimental data showing parallel operation of vacuum interrupters

Top two traces: currents, 1 div = 445A = 2,ms
Lowest trace: contact separation indicator - arrows show where contacts parted

(Courtesy of General Electric)

As noted in Chapter 2, a diffuse vacuum arc comprises a number of parallel arcs on the same pair of contacts. Each emanates from its own cathode spot and typically carries of the order of 100 A on copper. The extinguishing of spots by the transfer of current to other spots is no doubt dependent on the voltages of the individual arcs. As noted already there appears to be no limit to the amount of current one can carry and interruption in a diffuse vacuum arc; the contact surface area must be sufficient to accommodate the large number of parallel arclets required.

One might suppose that this notion could be extended to interrupters, i.e. to interrupt more current, put more interrupters in parallel. However, there are problems with this idea. If contact opening is not simultaneous, and if the spread straddles a current zero, those interrupters that are still closed after the current zero will carry all the current. They could be the majority or a small minority.

Likewise, if all parallel interrupters quenched their currents simultaneously at the same current zero, but one of the interrupters subsequently reignited, the operation would be a failure for the reignited interrupter and would find itself with the entire current. Clearly, this approach must be applied with caution. It might be possible for example, to counter such events as have just been described by a rapid close/open operation.

An interesting application for parallel operation of vacuum interrupters is the *generator switch*, which is used in some applications in place of a generator breaker. Auxiliary systems in power plants need electrical energy to enable the plants to be started up. This start-up power is frequently obtained from the power system by way of generator step up transformer; refer to Figure 10.23.

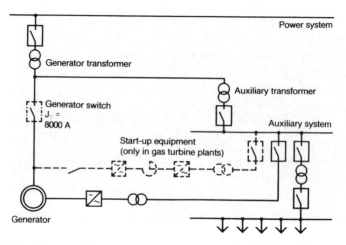

Figure 10.23   Circuit arrangement for starting generator from power system

This arrangement requires the ability to isolate the generator from its transformer. However, in some situations this does not call for a full-fledged generator breaker, a generator switch is used instead. Start-up is performed by way of the auxiliary transformer with the generator switch in the open position. The generator is taken to rated excitation and synchronised with the system by closing the generator switch. To shut down the generator, its load is reduced to zero, the generator switch is opened and the generator is de-excited.

It is clear that the generator switch must carry the full load current of the generator. It is for this reason that parallel interrupters are used. Figure 10.24 shows one embodiment. Each phase assembly comprises a three-pole, truck-mounted, medium voltage vacuum breaker, the three poles of which are connected in parallel to give the required current rating, which is 8000 A for this particular apparatus.

Figure 10.24   *Medium-voltage vacuum circuit breakers with their three poles connected in parallel to serve as individual poles for three phases of generator switch*

(Courtesy of Siemens AG)

A three-phase installation is shown in Figure 10.25. Note how the arrangement ensures phase isolation of the generator leads. It is an excellent example of what in Chapter 7 is called design for versatility.

10.13.3   Application in fusion machines

An application closely akin to the HVDC breaker is to be found in certain fusion machines. The objective is to transfer energy stored inductively in a primary winding, to a magnetically coupled plasma. This is achieved by using a forced-commutated vacuum circuit breaker to interrupt the current in the primary coil. Some thousands of operations at a substantial current are required, which makes vacuum an attractive candidate. Greenwood *et al.* [28] described such an application in 1972. The apparatus adapted a standard three-phase AC circuit breaker by adding a commutating circuit and surge suppressors. They also exercised the option of parallel operation by connecting the three poles of the circuit breaker in parallel. The equipment operated successfully for a protracted period on the Alcator project at the National Magnet Lab. (MIT). Benfatto *et al.* [33] used axial magnetic field interrupters for a similar purpose. Using two interrupters in series, they were able to interrupt 50 kA at 35 kV over 1500 times.

An even bigger fusion switching device has been described by Tamura *et al.* [34] which comprised four parallel paths for current, each path having two interrupters in series. The four sets of interrupters were arranged in a circle and

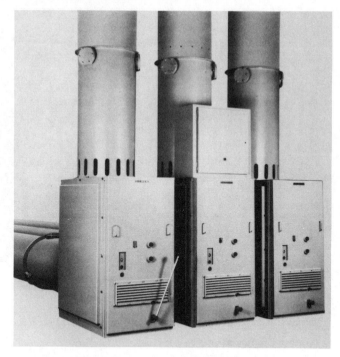

Figure 10.25   Three-pole generator switch rated 8000 A carrying
(Courtesy of Siemens)

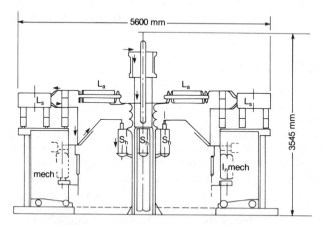

Figure 10.26   Drawing of DC switch which successfully interrupted 130 kA at 44kV
Current flow is shown by arrows

I_c = *vacuum valve*
L_a = *adjustable reactor*
S_h = *coaxial current shunt*
Mech = *actuating mechanism*
L_s = *saturable reactor*

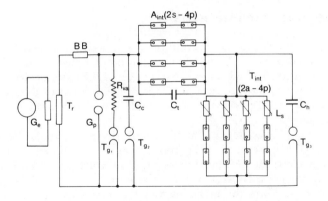

Figure 10.27   Circuit for testing switch shown in Figure 10.32 [34]

(*By permission of IEEE*)

G_e = *short circuit generator*
T_r = *short circuit transformer*
BB = *back up breaker*
G_p = *surge protection gap*
R_{va} = *energy absorbing resistor (0.25Ω)*
C_c = *commutating capacitor*
$T_{g1, 2, 3}$ = *trigger gap*

A_{int} = *auxiliary vacuum valves (25-4P)*
T_{int} = *test vacuum valves (25-4P)*
C_t = *voltage transfer capacitor (10 μF)*
C_h = *high voltage capacitor (150 μF)*
C_c = *1100 μF*
L_s = *(unsaturated)* 40 μH
      (saturated) 16 μH

fed current to the center like the spokes of a wheel. Figure 10.26 gives and impression of the equipment. It was tested on a 10 Hz AC source in the circuit shown in Figure 10.27.

10.13.4   Switching operations at nonstandard frequencies

It is common practice in power systems to switch at 50 or 60 Hz. However, sometimes applications arise at other frequencies. It has been noted already in Chapter 1 that vacuum switching devices were adopted quite early for high frequency applications, such as the switching of antennas. This remains a perfectly viable use. It is possible because of the extraordinary speed of which a vacuum switch recovers after arc extinction.

Successful operation is entirely possible at the other end of the frequency spectrum. The guiding consideration here is thermal. In general, lower frequencies imply longer arcing times inasmuch as the time between current zeros increases with decreasing frequency. As a rough rule of thumb the magnitude of the current that can be successfully switched can be expected to vary directly with the frequency. Thus, if an interrupter can switch 10 kA at 50 or 60 Hz, it should comfortably switch 5 kA at 25 or 30 Hz, since the thermal loading of the device is approximately the same. This 'rule' surely applies when the arc is diffuse. This is not to say, however, that this same device would interrupt 100 kA at 500 or 600 Hz, since at higher frequencies the speed of opening will be a factor.

10.14 Triggered vacuum gaps*

In a number of branches of engineering there is a need for a 'switch' that can close rapidly. Traditionally, this function has been performed by such devices as ignitrons or thyratrons and more recently by solid-state devices such as thyristors. In general, these devices are limited in the speed of their operation or else in the amount of power they can handle. An attractive alternate is the triggered vacuum gap (TVG).

TVGs have much in common with other multielectrode switches bearing the generic name of trigatron, in that conduction is established between the principal pair of electrodes by applying an impulsive voltage to a third electrode. This creates a discharge between the third electrode and one of the principal electrodes, which initiates the breakdown of the main gap. It can be shown, however, that there are some significant differences in performance between the TVG and conventional gas trigatrons.

Descriptions of triggered vacuum gaps [35–38] first appeared in the literature in the 1950s. For the most part, these were continuously pumped devices, since the technology did not exist for the construction of sealed, high-vacuum, high-power devices with a long life. The ability to produce virtually gas-free contact materials, most notably through the zone-refining technique, had a significant impact on this technology; as a consequence, it has developed more or less in parallel with the vacuum interrupter technology with which it has so much in common. Lafferty [39] gave a good account of this development and described some typical TVGs. TVGs and vacuum interrupters are very similar both in outside appearance and in construction. TVGs have a pair of principal electrodes separated by a gap like the contacts of a switch, except that neither electrode moves; that is, the gap is fixed. The trigger is located within one of the electrodes, usually on the axis. The size and precise design of a TVG - its rated current, voltage, and conduction time, - vary from application to application. Figure 10.28 illustrates two different designs. Note that one of these has two triggers; further comments are offered on this later.

The function of the trigger is to inject a plasma into the vacuum gap, this being the most effective way of producing a rapid breakdown with a minimum of jitter. This operation can be accomplished in several ways. One very convenient method that gives satisfactory performance involves using gas-loaded electrodes. Hydrogen has been used almost exclusively for the gas because of the ease of loading and the rapidity with which it is released from the electrodes on arcing. Only minute quantities of hydrogen are released, with no resultant buildup of hydrogen pressure on repeated operations of the main gap. After the plasma has been produced by the auxiliary discharge, it may be driven at high velocity into the main vacuum gap by a magnetic field. This field may be supplied externally or may result from the auxiliary discharge current itself. An example of the last type is described.

An arc may be initiated (through the application of a high-voltage pulse) between two hydrogen-loaded titanium electrodes that are separated by a few mils in vacuum; however the voltage must be of the order of tens of kilovolts to

* The material of this and the next section was prepared by the author for the treatise 'Vacuum arcs – theory and application,' coauthors J. D. Cobine, G. Ecker, G. A. Farrall and L. P. Harris, editor J. M. Lafferty. Copyright 1980 by John Wiley & Sons, Inc., whose permission to reprint the material is gratefully acknowledged.

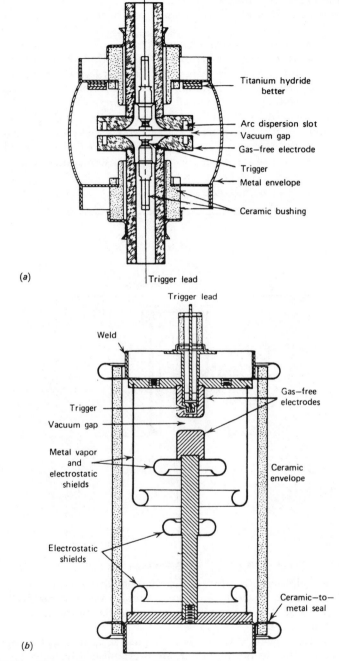

Titanium hydride
better

Arc dispersion slot

Vacuum gap

Gas—free electrode

Trigger

Metal envelope

Ceramic bushing

(*a*)

Trigger lead

Trigger lead

Weld

Trigger

Vacuum gap

Metal vapor
and
electrostatic
shields

Electrostatic
shields

Gas—free
electrodes

Ceramic
envelope

Ceramic—to—
metal seal

(*b*)

Figure 10.28   Triggered vacuum gaps [39]

  (*a*) *Cross-sectional view of TVG for 60 Hz operation*
  (*b*) *High-voltage TVG for microsecond pulse operation*

give reliable breakdown that is free from jitter. If an insulator is placed in the space between the electrodes, breakdown occurs over the insulator surface at a lower voltage than in the simple vacuum gap, and breakdown time and jitter may be kept small with the application of only a few kilovolts. Figure 10.29 shows one such arrangement. A high-alumina ceramic cylinder is coated with a titanium layer only a few mils thick. Other metals forming hydrides, such as hafnium, palladium, thorium, uranium, yttrium, zirconium, and the rare earths, can be used depending on the method of processing of the gap and operating temperatures involved. A V-notched groove is then cut through the titanium into the ceramic. A shield cap is placed on one end of the ceramic cylinder, and a lead wire is attached to the cap. This assembly is inserted into a conical recess in the cathode electrode of the main vacuum gap as shown in Figure 10.29. Gas

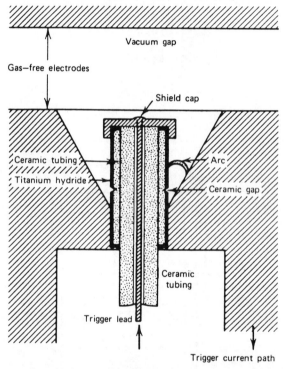

Figure 10.29   Cross-sectional view of arrangement for triggering vacuum gap [39]

loading is accomplished by heating the titanium in hydrogen. In alternative designs, the trigger is loaded before final assembly or it is loaded by heating in hydrogen with a built-in heater after assembly; or the entire cap assembly may be heated in hydrogen.

To operate the trigger, a positive voltage pulse is applied to the trigger lead. The ceramic gap breaks down, and an arc is established between the titanium hydride electrodes, thus releasing hydrogen and titanium vapour, which are ionised and sustain the discharge. Expansion and magnetic forces produced by the discharge current loop drive the plasma out of the conical recess into the main gap. As the plasma spreads out into this gap, a high-current glow discharge is first established between the main electrodes. The glow is transformed into an arc as cathode spots are established on the negative main electrode with release of metal vapor. Measurements indicate that with peak current pulses of 10 A on the trigger electrode, the main gap will break down in less than 100 ns, with jitter times of about 30 ns, when 30 kV is applied to the main gap. The trigger energy required is less than 0.01 J. The main gap may be broken down with a trigger voltage pulse as low as 50 V, however longer delay times result.

A triggered vacuum gap can be made to operate if the polarity of the main electrodes is reversed, that is, if the trigger is located in the electrode is momentarily the anode. The triggering effort required in this situation is significantly higher. For this reason a trigger is sometimes put in both the principal electrodes, as in Figure 10.28(a), when the device is to be operated in an AC circuit with random polarity.

Once a TVG has been triggered, it behaves very much like a conventional vacuum interrupter; for example, when the current it is carrying passes through zero, it interrupts. It can carry tens of thousands of amperes with an arc drop of a few hundred volts or less; but after interruption the metal vapour supplied by the electrodes condenses on the shield and electrodes, permitting the gap to recover its full dielectric strength in a few microseconds, as would a vacuum interrupter.

A remarkable feature of the TVG is the extraordinary wide range of operating voltages. A single gap with a 0.086 inch spacing was found to operate from 300 to 30000 V. Below 1 kV the delay time was of the order of 1 μs, and above 1 kV the delay was less than 0.1 μs.

The triggered vacuum gap is more expensive than the conventional three-electrode spark gap operating in air, but it has a number of advantages, including

- small size
- rapid deionisation
- operation possible over a wide range of gap voltage
- operation possible in a strong radiation environment
- gas cleanup presents no problem
- no audible noise or shock waves
- no explosion hazard.

The applications for TVGs are many and varied. One group can be categorised as high-speed protective devices that operate with extreme rapidity to short out a component in danger of damage by overvoltage or overcurrent. Equipment to be protected has included high-power klystrons and other electronic components; in power circuits TVGs can be used to protect series capacitors in high-voltage transmission lines. Under fault conditions it is essential that the capacitors be temporarily shorted out to avoid subjection to overvoltage.

In Figure 4.36, illustrating the use of vacuum interrupters for the interruption of high-voltage direct current, a switch S is indicated in the commutating circuit. This is a good application for a TVG. The use of a similar circuit for interrupting the direct current in the ohmic heating coils of Tokamaks also was described earlier; this again offers an opportunity for the TVG as the instrument to complete the commutating circuit. There are a number of other applications for 'make-switches' in the fusion area.

K.D. Ware *et al.* [41] described the design and operational characteristic of a TVG and its mechanical attachment to two kinds of industrial capacitor, used singly or in multicapacitor modules. The main features of the vacuum device are its wide range of voltage operation (0.1–50 kV) with relatively easy triggering, high-current capability (approximately 400 kA), low inductance (approximately 5 nH), and long life. High fidelity with minimum jitter time of a few nanoseconds was claimed.

A quite different application is described by J.M. Anderson [42]. Here a TVG was used to switch microwaves in evacuated wave guides.

An application in the power area, described by Lee and Porter [43], relates to equipment that provides precise control in the initiation of very high currents required for testing circuit breakers and other power system components. The TVG used for this current initiation is quickly shunted by parallel electromechanical switch for continuous conduction, a useful approach where prolonged conduction is required. It avoids the very serious erosion, therefore loss of life, that would occur if the TVG carried the current by itself for the full duration, and also the necessity to trigger the TVG on successive halfcycles.

The hybrid approach, designated 'triggered vacuum interrupter' (TVI), is an interesting alternative for the above-mentioned application. The TVI can be looked on as a vacuum interrupter with a trigger in its fixed contact, or alternatively as a TVG with a movable electrode. Such a device can precisely and quickly complete a circuit by firing its trigger; subsequently the contacts can be closed to carry the current through a solid metallic connection. The current can be interrupted in an AC circuit by opening the contacts again, since the device then behaves like a vacuum interrupter.

Since regular TVGs have a permanent gap, they are less constrained as far as contact materials are concerned than are vacuum interrupters; for instance, there is no constraint on welding. A TVI, on the other hand, is subjected to the constraints of vacuum interrupters and TVGs when a contact material selection is made. Recently there has been a trend away from metal-hydride triggers. Successful TVGs have been developed with metal or metal-film triggers [44–46].

10.15 Vacuum fuses

Vacuum fuses are yet another example of the use of high vacuum for the current interruption function [47]. Like TVGs they have a pair of opposed electrodes that typically are similar in form to the contacts of a vacuum interrupter. However, these fixed electrodes are bridged by a fuse element that conducts current under normal conditions. On the incidence of a fault or other severe abnormal condition, this element 'blows'; that is, it melts and vaporises, creating an arc in the metal vapour so produced. Thereafter the device functions much like an arcing vacuum interrupter with its contacts in the open position.

The critical component in this assembly is the fuse element itself. Like the contacts of a vacuum interrupter, it must be made from virtually gas-free material so that the arc, when it develops, is sustained in a condensable medium. For this reason the bridging element is prepared with at least the same care as the electrodes themselves.

A traditional problem associated with fuses is posed by conflicting

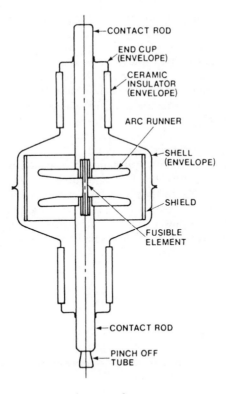

CONTACT ROD

END CUP
(ENVELOPE)

CERAMIC
INSULATOR
(ENVELOPE)

ARC RUNNER

SHELL
(ENVELOPE)

SHIELD

FUSIBLE
ELEMENT

CONTACT ROD

PINCH OFF
TUBE

Figure 10.30   Schematic diagram of vacuum fuse components

requirements of current carrying and current interruption. For interruption one would wish the process to be extremely rapid with severe faults. More time can be tolerated for less severe faults; indeed, the device must be able to discriminate between low-grade faults on the one hand and normal overloads and temporary current surges, such as accompany the starting of a large motor, on the other. This leads to the concept of an inverse time characteristic by which the speed of the fault is inversely proportional to the magnitude of the fault current.

As load current requirements increase, this type of characteristic is ever more difficult to obtain; a rugged fuse element capable of carrying the load current and associated normal overloads will fail to blow or will blow too slowly when a fault occurs; whereas a lighter fuse element that can respond quickly to a fault may melt and vaporise when this behaviour is not required with the advent of an acceptable overload condition.

It is claimed that the vacuum fuse has superior performance in this regard. The relatively short physical length of the fuse element allows the heat generated in it by the normal load current to be conducted out through the main contacts. These contacts are also capable of storing or dissipating the heat transferred to them under temporary overload conditions. Under fault conditions the heat cannot escape quickly enough, and the fuse melts at a neck in its geometry and creates an arc. The added heat generated by the arc rapidly vaporises the remainder of the fuse element, creating conditions essentially identical to those in a vacuum interrupter prior to interruption, and the current is ruptured as a consequence.

The vacuum fuse has many of the characteristics of the vacuum interrupter because it has many of its physical attributes; this is evident from Figure 10.30. Both 'contacts' are fixed; thus the device has no bellows and of course requires no mechanisms. But in other respects the vacuum fuse and the vacuum interrupter are quite similar. Therefore vacuum fuses are relatively sophisticated devices and relatively expensive. However, the performance they offer has rendered them quite attractive for certain applications, most notably as high-speed sectionalisers in 15 and 34.5 kV power distribution circuits.

An advantage claimed for the vacuum fuses is that no gas is expelled from the device when it operates since the fuse element is totally enclosed within the vacuum envelope. This means that such fuses can be used under oil, or where used in air, smaller clearances can be used to energise portions of the circuit. Unlike some other high-power, high-voltage fuses, this device is quiet in operation; there is no impact. Also, it is small and lightweight compared with many other fuses. A 15 kV, 300 A vacuum fuse, capable of interrupting 12,000 A, weighs approximately 2.5lb.

10.16   References

1 GREENWOOD, A.: 'Electrical transients in power systems' (Wiley, New York, 1991, 2nd edn.) pp. 265–271
2 Reference 1, pp. 51, 52
3 Reference 1, p.83
4 HARNER, R.H., and RODRIGUEZ, J.: 'Transient recovery voltages associated with power system three-phase transformer secondary faults', *IEEE Trans.*, 1972, **PAS–91**, pp. 1887–1896
5 'Small inductive current switching' CIGRE WG 13.02, Chap 1–4, *Electra* (72, 75, 101 and 112)
6 LEE, T.H.: 'The effect of current chopping in circuit breakers on networks, and transformers, part I, theoretical considerations', *Trans. AIEE*, 1960, **79**, p. 535
7 GREENWOOD, A.N., KURTZ, D.R., and SOFIANEK, J.C.: 'A guide to the application of vacuum circuit breakers', *IEEE Trans.*, 1971, **PAS–90**, pp.1589–1597
8 KANEKO, F., YOKOKURA, K., OKAWA, M., and OHSHIMA, T.: 'Possibility of high current interruption of vacuum interrupters with low surge contact materials: improved Ag/WC', IEEE Winter Power Meeting, 1992, New York, USA, paper 92WM 260-0-PWRD
9 Reference 1, pp 327–331
10 IEEE/ANSI standard C57.12.14: 'Dielectric test requirements for power transformers for operating at system voltages from 115 kV through 230 kV', 1982
11 ITOH, T., MURAI, Y., OHKURA, T., and TAKAMI, T.: 'Voltage escalation in switching of motor control circuit by vacuum contactor', *IEEE Trans.*, 1972, **PAS–91**, pp. 1897–1903
12 GREENWOOD, A., and GLINKOWSKI, M.: 'Voltage escalation in vacuum switching operations', *IEEE Trans.*, **PD–3**, 1988, pp. 1698–1706
13 GREENWOOD, A.: Discussion of MURANO, M. FUGII, T., NISHIKAWA, H., and NISHIKAWI, S.: 'Voltage escalation in interrupting inductive current by vacuum switches' and 'Three-phase simultaneous interruption in interrupting inductive current', *IEEE Trans.*, 1974, **PAS-93**, p. 278
14 OSMOCROVIC, P.: 'Influence of switching operations on the vacuum interrupter dielectric strength', *IEEE Trans.*, 1974, **PWDR–6**, pp. 1073–1080

15 MORRISON, R.: 'Grounding and shielding techniques in instrumentation' (Wiley, New York, 1986, 3rd edn.)

16 O'LEARY, R.P., and HARNER, R.H.: 'Evaluation of methods for controlling the overvoltages produced by energisation of shunt capacitor banks', CIGRE report 13-05, 1988

17 GREENWOOD, A.: 'Electrical transients in power systems' (Wiley, New York, 1991, 2nd edn.) pp. 132–138

18 'Schedules of preferred ratings and related required capabilities for AC high voltage circuit breakers rated on a symmetrical current basis', ANSI standard C37.06-1979

19 'Application guide for AC high voltage circuit breakers', ANSI standard C37.010, 1979

20 CORNICK, K.J., and THOMPSON, T.R.: 'Steep-fronted switching voltage transients and their distribution in electrical machine windings, part 2: distribution of steep-front voltage transients in motor windings', *IEE Proc. B*, 1982, **130**, p. 56

21 ORACE, H., and McLAREN, P.G.: 'Surge voltage distribution in line-end turns of induction motors', *IEEE Trans.*, 1985, **PAS–104**, pp. 1843–1848

22 RECKLEFF, J.G., NELSON, J.K., MUSIL, R.J., and WENGER, S.: 'Characterisation of fast rise-time transients when energising large 13.2 kV motors', *IEEE Trans.*, 1988, **PD–3**, pp. 627–636

23 DICK, E.P., GUPTA, B.K., PILLAI, P.R., NARANG, A., LAUBER, T.S., and SHARMA, D.K.: 'Prestriking voltages associated with motor breaker closing', *IEEE Trans.*, 1988, **EC–3**, pp. 855–863

24 GUPTA, B.K., DICK, E.P., GREENWOOD, A.N., KURTZ, M., LAUBER, T.S., LLOYD, B.A., NARANG, A., PILLAI, P.R., and STONE, G.C.: 'Turn insulation capability of large AC motors, vols. 1 and 2', EPRI report EL-5862, 1988

25 KEERTHIPALA, W.W.L., McLAREN, P.G., and GREENWOOD, A.N.: 'Some factors governing the severity of swithcing surges', Proceedings of IEEE IAS annual meeting, 1989, San Diego, CA, USA, pp.1886–1889

26 HARDER, J.E.: 'Metal oxide arrester ratings for rotating machine protection', *IEEE Trans.*, 1985, **PAS–104**, pp. 2445–2452

27 McLAREN, P.G., and ABDEL-RAHMAN, M.H.: 'Steep-fronted surges applied to large AC motors—effect of surge capacitor value and lead length',*IEEE Trans.*, 1988, **PWRD–3**, pp. 990–997

28 GREENWOOD, A.N., BARKAN, P., and KRACHT, W.C.: 'HVDC vacuum circuit breakers', *IEEE Trans.*, 1972, **PAS–91**, pp. 1570–1574

29 YANABU, S., TAMAGAWA, T., IRAKAWA, S., HORIUCHI, T., and TOMIMURO, S.: 'Developemnt of HVDC circuit breaker and its interrupting test', *IEEE Trans.*, 1982, **PAS–91**, pp. 1575–1588

30 GREENWOOD, A.N., and LEE, T.H.: 'Theory and application of HVDC circuit breakers', *IEEE Trans.*, 1972, **PAS–91**, pp. 1575–1588

31 'The metallic return transfer breaker in high voltage direct current transmission', CIGRE WG13.03, *Electra* (68), 1980, pp. 21–30

32 BACHMANN, B., MAUTHE, G., RUOSS, E., LIPS, H.P., PORTER, J., and VITHAYATHIL, J.: 'Development of a 500 kV airblast HVDC circuit breaker', *IEEE Trans.*, 1985, **PAS–104**, pp. 2460–2466

33 BENFATTO, T., DeLORENZI, A., MASCHIO, A., WEIGAND, W., TIMMERT, H.P., and WEYER, H.: 'Life tests on vacuum switches breaking 50 kA unidirectional current', *IEEE Trans.*, 1991, **PWRD–6**, pp. 823–832

34 TAMARU, S., SHIMADA, R., KITO, Y., KANAI, Y., KOIKE, H., IKEDA, H., and YANABU, S.: 'Parallel interruption of heavy direct current by vacuum circuit breakers', *IEEE Trans.*, 1980, **PAS–99**, pp. 1119–1129

35 HAGGERMAN, D.C., and WILLIAMS, A.H.: 'High power vacuum spark gap', *Rev. Sci. Instrum.*, 1959, **30**, pp. 182–183

36 BAKER, W.R.: 'High voltage, low inductance switch for mega-amperes pulse current', *Rev. Sci. Instrum.*, 1959, **30**, pp. 700–702

37 BRACEWELL, G.M., MAYCOCK, J., and BLACKWELL, G.R.: 'Switching two million amperes', *Nucl. Power* GB, 1959, **4**, 115–117

38 BRISH , A.A. *et al.*: 'Vacuum spark relays', *Instrum. Exp. Tech.*, 1958, **6**, pp. 644–649

39 LAFFERTY, J.M.: 'Triggered vacuum gaps', *Proc. IEEE*, 1966, **54**, (1), pp. 23–32

40 GLEICHAUF, D.H.: 'Electrical breakdown over insulawtion in high vacuum', *J. Appl. Phys.*, 1951, **22**, pp. 535–541
41 WARE, K.D. *et al.*: 'Design and operation of a fast high-voltage vacuum switch', *Rev. Sci. Instrum.*, 1971, **42**, pp. 512–518
42 ANDERSON, J.M.: 'Microwave switching in evacuated waveguides by metal vapor arcs', *IEEE Trans.*, 1970, **ED–17**, (10), pp. 939–940
43 LEE, T.H., and PORTER, J. W.: 'A closing switch shunted by a triggered vacuum gap for precise current initiation', Proceedings of the international symposium on High Power Testing, Portland, OR, USA, July 1971
44 LAFFERTY, J.M.: 'Vacuum gap device with metal ionizable species evolving triggering assembly', US patent 3 465 205, 1969
45 GOODY, C.P.: 'Coaxial electric arc discharge devices', US Patent 3 719 852, 1973
46 BOXMAN, R.L.: 'Trigger mechanism in triggered vacuum gaps', *IEEE Trans.*, 1977, **ED–24**, pp. 122–128
47 MAREK, J.R.: 'Vacuum fuses—an innovation in circuit protection', Proceedings of American Power Conference 36, 1974, pp. 965–970

Chapter 11

Testing vacuum switchgear

11.1 Introduction

A piece of switchgear, be it a circuit breaker, recloser, load-breaks switch or contactor has a certain rating with respect to voltage and current. It is built to conform with particular standards as explained in Section 10.2 where standards are discussed. It is the manufacturer's business to see that the product meets its rating by performing tests in accordance with the standards and having the results certified by an independent testing facility. The user, at his option and expense, may request that some of these tests be repeated and that a representative witness the tests. The manufacturer will continue to test proven products from time to time as part of his quality assurance. These types of tests are performed regardless of the switching technology involved and are not discussed in this chapter. Instead, it concentrates on testing that is peculiar to vacuum and where the test must be modified because the technology is vacuum, we will tend to look at things from the user's point of view, and answer questions that are of most concern the user.

11.2 Testing for vacuum

'How do I know I still have vacuum in an interrupter?' This so familiar question is the first that comes up in almost every discussion the author has had on vacuum for the past 40 years. In the course of manufacture the vacuum integrity of welds and brazes are checked very thoroughly as noted in Section 9.2.5, but the helium leak checking can only be done on subassemblies or before the entire interrupter is sealed off, it is not an available option for the field. Nor are magnetron pressure measuring devices typically available to the user. It is quite straightforward to determine whether an interrupter has lost its vacuum completely, if one is holding the component in one's hands. The contacts of a healthy interrupter are forcibly held together by the atmospheric pressure acting on the bellows; it requires quite an effort to pull the contacts apart. If vacuum has been lost, the moving contact is slack and sloppy. Switches and breakers are designed to make such a check with the interrupters *in situ*. It is a matter of disengaging the springs that hold the contacts closed.

A partial loss of vacuum requires a different means of detection; the normal method is by hipot which essentially interrogates the situation through the Paschen curve (Figure 3.3). A typical test procedure for all voltage ratings through 15 kV max calls for a power-frequency voltage of 36 kV RMS to be applied across the open contacts for one minute. Successful withstand indicates that the interrupter is satisfactory. To avoid any ambiguity in the AC high potential test due to leakage or displacement current, the test unit should have sufficient volt–ampere capacity to deliver 25 mA for one minute.

X-radiation produced during this test with recommended voltage and normal contact spacing is extremely low and well within maximum permitted by standards. However, as a precautionary measure against the possibility of applying higher than recommended voltage and/or less than normal contact spacing, it is recommended that all personnel stand at least one metre away from the breaker.

A DC hipot test is an acceptable alternative, the comparable test voltage is 40 kV which must be held for one minute. The test equipment should be capable of delivering 5 mA for one minute to avoid ambiguity due to field emission or leakage currents.

Concern for loss of vacuum is understandable but really not justified in light of the record. The incidence of this problem is minimal and serious consequences of it are minuscule.

11.3 Non-sustained disruptive discharge

In the course of testing a vacuum switching device a non–sustained disruptive discharge may occur. This phenomenon, peculiar to vacuum, is discussed in Section 4.2.2. Such events have led to concern over the interpretation of IEC publication 56. The matter has been taken up by the Short-Circuit Testing Liaison (STL), which has the following to say on the subject in its Guide to Interpretation of IEC publication 56 [1].

'On occasions, one or more non–sustained disruptive discharges may occur during the recovery voltage period following a breaking operation. A non-sustained disruptive discharge is defined as a disruptive discharge between the contacts during the power frequency recovery voltage period resulting in a high-frequency current flow which is related to stray capacitance local to the interrupter. The presence of such a discharge can be recognised by a study of the test oscillogram and is characterised by the representations in Figures 11.1 and 11.2.

Figure 11.1 represents a typical three-phase characteristic and shows a momentary collapse of the recovery on one of the phases which may result in a corresponding offset of the power-frequency recovery voltage waveforms on all phases, the amplitude and time constant of which is dependent on the local values of L, R and C. Figure 11.2 shows the equivalent single-phase characteristic.

Sub-clause 6.102.7 states that during making and breaking tests, the circuit breaker shall neither show signs of excessive distress nor endanger the operator. The occurrence of a non-sustained disruptive discharge is interpreted as showing some signs of distress. It is the degree to which such distress is observed that is of importance and to this end, the following conditions apply for the acceptance of a circuit breaker for certification, in the event of such a phenomenon occurring.

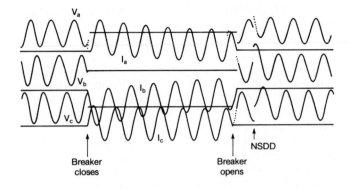

Figure 11.1   *Typical non-sustained disruptive discharge in three-phase circuit*

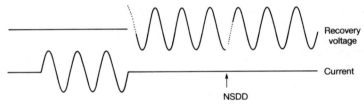

Figure 11.2   *Typical non-sustained disruptive discharge in single-phase circuit*

(1) The power-frequency recovery voltage period shall be extended to at least 0.3 s after current interruption irrespective of the phenomenon occurring or not. Other conditions of Clause 6.104.7 being applicable.

(2) Certification will not be allowed if there are four or more occurrences of the phenomenon during an entire series of test duties required by the Standard including any test duties which are repeated or restarted for whatever reasons. When there are multiple occurrences of the phenomenon during the recovery voltage period of one test they shall be counted individually.

(3) Resumption of power frequency related follow-through current triggered by a non-sustained disruptive discharge is not allowed even though it may result in a single loop of current to flow.

Where a Certificate is granted for a circuit breaker based on a series of tests during which a non-sustained disruptive discharge occurred, a statement shall be included within the Certificate as follows:

'During the tests forming this Certificate, (a) non-sustained disruptive discharges occurred during the recovery voltage period on test duty(ies) . . .'

It is the author's view that this guideline is discriminatory. Just because vacuum has the unique ability to burp on occasions to relieve its indigestion, it should not be punished, indeed, it should be commended.

11.4   Testing for contact condition after undergoing short-circuit

Standards require that when a circuit breaker has undergone a short-circuit test, the condition of the contacts should be such that it can carry its normal rated current without exceeding thermal limits. Traditionally, a determination was made either by inspection or by a heat run. It is clearly impossible to inspect the contacts of a completely sealed device, it therefore seemed that the matter must be settled by heat run.

This decision precipitated a difference of opinion, vacuum breaker manufacturers asserting that requiring vacuum interrupters to conform to the temperature rise prescribed for other technologies was unreasonable inasmuch as the pristine conditions within the vacuum envelope allowed the contacts to be operated at a higher temperature without ill effects. This state of affairs was mentioned in Section 5.7.3.

A compromise was reached by the SIL [1] which states 'The condition of the contacts after the short-circuit tests is considered acceptable if the maximum temperature rise recorded at the terminals of any vacuum interrupter does not exceed by more than $10°K$ the values specified in Table V of IEC publication 694. The temperature rise values stated for connections, bolted or equivalent, are applicable'.

11.5   Determining switch life

Long life is an attractive feature of vacuum interrupters, however, their life is not infinite. Under normal circumstances the limiting components are the contacts. Whenever a switch arcs, some contact material is boiled off the contact surface. The quantity is minute for load and no-load operations, but is much more substantial when a short-circuit current is being interrupted, or if the switch closes into a short circuit, especially if the arc is constricted (Section 2.2) when gross melting of the contacts can occur causing contact material to be removed as molten droplets. Thus, in course of time, the contacts are sufficiently eroded to require replacement of the interrupters.

The contacts are held together forcibly by the wipe springs as explained in Section 6.2, and inasmuch a these springs are precharged, the contact force does not change significantly when erosion occurs. As the erosion proceeds, a discernable change can be observed in the closed position of the moving contact; it advances into the interrupter. This is apparent from a sink mark on the shank of the moving contact or its support. An example of such an indicator is shown in Figure 11.3. The position of the indicator (28) relative to the stripe (28.1) indicates changes in the contact system. The breaker may be operated only when the indicator is within the range of the stripe. A somewhat similar arrangement is to be observed in Figure 6.15.

In normal duty an interrupter rarely has to be changed because of excessive erosion. Frequent, heavy switching, as can be experienced with arc furnaces, may require such a changeout.

Figure 11.3   Erosion check marking on vacuum circuit breaker
(Courtesy Siemens AG)

11.6   Testing for X–rays

At several points in this text reference has been to the production of X–rays (Section 11.2 of this Chapter, for example). This is because the normal way of producing x-rays is to impress a high voltage across closely spaced electrodes in vacuum. *However*, the electric field present in a vacuum interrupter during *normal* operation is quite inadequate to produce any hazard to personal from X-rays. In spite of this, there is a standard which addresses the subject [2]. It sets limits for the amount of radiation permitted and describes conformance tests for new interrupters to check on their compliances. For interrupters in service it specifies 75% of the 'as new' voltage when performing dielectric withstand tests (much along the lines of what was described in Section 11.2 for checking vacuum) to avoid any danger from X-rays. It seems probable that this standard's main purpose is safeguard against possible legal actions.

X-rays may become a more important issue in the future if high voltage interrupters, operating at higher electric field strength, are developed.

As a sideline to this discussion we would mention a potential standard C37.59-199X 'Requirements for conversion of power switchgear equipment', which is in the approval process at the time of writing. This document addresses the conversion of air magnetic circuit breakers to vacuum. There is considerable concern in this area of improper application of vacuum interrupters based on the required contact force, velocity, and possibility of contact bounce*. Further, there is the legal implications of the name plate from the original manufacture as well as the responsibility of the company that undertook the conversion.

---

* E. Vevierka, private communication

11.7   Comparative testing of vacuum switchgear products

11.7.1   Problems of testing

A general problem of switchgear testing is that it is expensive because of the high capital cost of traditional testing equipment and the high cost of operating and maintaining this equipment.

Synthetic testing is an alternate to direct testing. In such a test the device being tested is made to 'believe' it is experiencing the conditions of the system environment in that the current, short circuit or otherwise, corresponds to field condition, and the TRV impressed across the interrupter following current zero is also appropriate to what it would experience in the power system, although in fact the current and voltage are being supplied from separate and distinct sources in synthetic testing. This approach can often be less expensive, but its main advantage is that it permits testing at power levels well beyond the direct testing capacity of the largest test station.

What is about to be described is a synthetic method which specifically capitalises on a certain property of vacuum interrupters, namely, their ability to interrupt high-frequency currents. It is illustrated by applying it to capacitance switching, which is where it was proposed and used by Glinkowski *et al.* [3].

The stresses experienced by a circuit breaker while disconnecting a capacitor bank are described in Section 10.6.3. After current zero a $(1-\text{cosine})$ recovery voltage at power frequency, is impressed between the contacts as indicated in Figure 10.13. Capacitance switching is a frequent operation since capacitors are added or removed in response to daily load cycles to reduce voltage fluctuations. Thus, a high integrity is required of breakers that perform the capacitance switching function. If a utility company has 50 capacitor bank installations that are switched twice a day, there are 36,500 operations a year. If a restrike occurs once in every 5000 operations, it means that there would be seven such incidents a year. To perform a normal direct test program to check on these kind of probabilities would require a quite unacceptable amount of testing. The method to be described is intended to address this specific problem.

11.7.2   Test method

Successful switching of any kind depends on the rapid recovery of dielectric strength between the switch contacts, following current interruption. A method is described in Section 4.2.2 for measuring this characteristic; the circuit is shown in Figure 4.12 and some typical results are to be seen in Figure 4.14, where we observe the breakdown voltage as a function of time after contact parting. The data of this last figure are for so-called cold gap conditions, that is with essentially no prior arcing. It will be seen that there is significant dispersion in the breakdown voltage values.

The test circuit for the present purpose is shown in Figure 11.4. Like many other synthetic test circuits, it has separate current and voltage sources. A preselected power frequency current, corresponding to the capacitor bank current being simulated, flows through the test breaker from the current source. The opening of the test breaker is precisely timed so that the period of arcing, from $0–60°$ ($0–2.78$ ms at 60 Hz), can be controlled to less than 100 μs. The

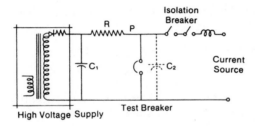

Figure 11.4   Test circuit for dielectric recovery evaluation following current switching [4]
(By permission of IEEE)

reason for this is as follows. The most highly stressed pole when switching a three-phase, ungrounded, capacitor bank is the first pole to clear. If arcing extends more than 60°, it is no longer the first pole to clear.

The isolation breaker divorces the current source from the remainder of the circuit following current interruption. Simultaneously, the local stray capacitance, designated C_2 in the diagram, charges through R, applying a rising voltage across the separating contacts of the test breaker. The source of this voltage is the 120 kV DC supply on the left in Figure 11.4. The form of the voltage is (1−exponential) with a time constant RC_2. In this respect, the test conditions are different from the power frequency (1−cosine) recovery voltage experienced by a circuit breaker in the field. However, by rapidly overstressing the contact gap, the method assures that breakdown will occur. Moreover, the discharge current precipitated by the breakdown is interrupted almost immediately because it is so small, whereupon the gap recovers and voltage climbs until another breakdown occurs in the manner shown in Figure 4.13. By this means, tens of breakdown can be observed in a single test, allowing much data on the dielectric recovery of the contact gap, following switching of the appropriate power-frequency current, to be collected relatively quickly. The isolation breaker is subjected to the same stress as the test breaker and must not break down. This is assured by using a three-phase breaker for the isolation function with its poles connected in series.

The pattern of breakdown observed with prior power frequency arcing is essentially identical with the cold gap data obtained by Roguski [4] and reproduced in Figure 4.13. As noted already, there is a wide dispersion of breakdown voltage values which suggests that the data are best treated statistically. How this is carried out is now described.

11.7.3   The Weibull distribution

The Weibull distribution has been used extensively in reliability studies when it was desired to determine the probability of a component surviving for a particular length of time, the failure rate of supposedly like components, etc. It appeared therefore that it might be a suitable candidate for establishing the probability of reignition during capacitance switching.

A continuous random variable V has a Weibull distribution if its density function is

$$f(V) = \alpha\beta V^{\beta-1} \alpha V^\beta, \quad V > 0$$
$$= 0, \text{ elsewhere}$$
(11.7.1)

α and β are positive parameters. The cumulative distribution of V, $F(V)$, the probability that V is less than some value V_a, say, is

$$F(V) = \int_0^{V_a} f(V)\,dV$$
(11.7.2)

Similarly

$$1 - F(V) = \int_{V_a}^\infty f(V)\,dV$$
(11.7.3)

A set of data has a Weibull distribution if, when

$$\ell n\,\ell n\left[\frac{1}{1 - F(V)}\right]$$

is plotted against $\ell n\,\ell n(V - V_{min})$, the result is a straight line. V_{min} is here the lowest recorded value of V in the set of data.

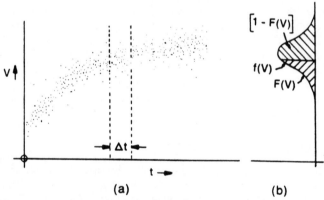

Figure 11.5   *Incidence of reignition as function of time after current zero for tests of particular arc duration*

(By permission of IEEE)

The variable in the present context is the voltage across the interrupter contacts at the instant of breakdown. There have been a number of previous attempts to determine the probability of breakdown from statistical data. Murai *et al.* [5] made tests somewhat similar to Roguski [4] but observed the breakdowns on closing. Later, the same group applied impulse voltages to a fixed gap [6]. More recently, Osmokrovic and Djogo [7, 8] made similar experiments. All these investigators used Weibull statistics to evaluate their results.

The procedure for predicting the failure rate at any particular voltage is as follows. Starting with a plot similar to Figure 4.14, containing all the data from tests made with a relatively narrow band of arcing times, the data for a particular time interval Δt is extracted (Figure 11.5(*a*)). The distribution density function, $f(V)$ for this population is shown in Figure 11.5(*b*). $F(V)$ and $[1\text{-}F(V)]$ are also indicated (by the shaded areas). The lowest value of reignition voltage is taken as V_{min}, the remainder are organised as $(V\text{-}V_{min})$ in ascending order. The corresponding values of $[1\text{-}F(V)]$ are computed and a Weibull plot of

$$\ell n \, \ell n \left[\frac{1}{1 - F(V)} \right]$$

against $\ell n (V - V_{min})$ constructed. As stated, if the data follows a Weibull distribution this plot will be a straight line. That this turned out to be the

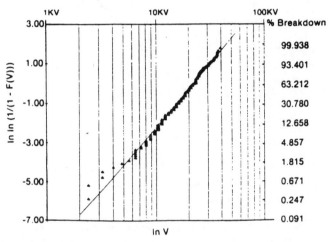

Figure 11.6   *Typical Weibull plot of reignition data [4]*

(*By permission of IEEE*)

case is evident from Figure 11.6 which shows a typical plot obtained in this manner. The straight line shown was arrived at by linear regression analysis. This line can be projected back to determine the probability of reignition at any other voltage. Of particular interest is the voltage that would be present at the instant corresponding to Δt during a capacitance switching operation. Figure 11.7 shows the (1–cosine) TRV.

The procedure is repeated for every time interval for which there is a significant amount of data. The probability of a reignition during capacitance switching for this particular range of arcing times is obtained by summing the probabilities for all time intervals Δt.

The entire procedure must be repeated for all the other bands of arcing time or arc angle in the range 0–60°, if it is observed that there is a dependence on

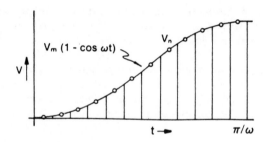

Figure 11.7   (1—cosine) recovery voltage experienced by circuit breaker when disconnecting capacitor bank [4]

(By permission of IEEE)

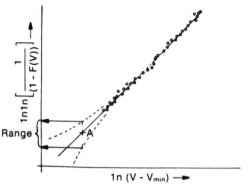

Figure 11.8   Extrapolating from Weibull plot [4]

(By permission of IEEE)

arcing time. Since all arc angles in the range are equally likely, the overall probability of a reignition on the first phase to clear is the average value.

Projecting the Weibull plot line to lower voltage levels does not in fact pinpoint the probability of reignition at another voltage. Rather, it defines a range within which, to a given degree of confidence, the probability will be. This is illustrated in Figure 11.8. The dotted lines define the error for a given degree of confidence. The range for the particular voltage *A* is as indicated. The more data there is available, the more closely will the dotted lines approach the straight line and the narrower will the range become. If a higher degree of confidence is demanded, the dotted lines will be forced apart. One could work with the probability corresponding to the top of the range. This would give a pessimistic, conservative prediction.

So far the possibility of reignition on the other two poles has been ignored. The probability of such an event is likely to be low since each sees less than 70% of the voltage experienced by the first pole to clear. Also, if it is assumed that the contacts of all three poles part in unison, the arc angles for poles 2 and 3 will be 90° longer than for pole 1, which means that their contact gaps will be appreciably longer by the time their TRVs are impressed across them. It is, of

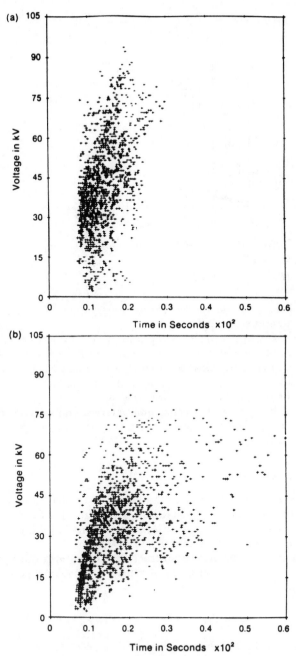

Figure 11.9   *Test data for arcing time of 750 μs [4]*

    (a)   *Cu/Cr*
    (b)   *Cu/Bi*

    (*By permission of IEEE*)

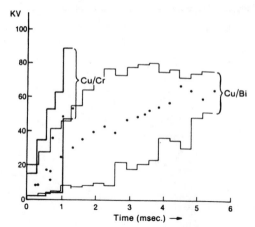

*Figure 11.10   Comparison of breakdown data for Cu/Cr and Cu/Bi interruption from short
arcing time tests [45]*

(By permission of IEEE)

course, possible to calculate the probability of reignition for poles 2 and 3 by the
same method described for pole 1.

Glinkowski *et al.* [4] developed and put in place a programme for processing
data obtained from tests. It progressed step by step through the Weibull analysis
described; the procedure was essentially automatic. Each reignition as it
occurred during a test was recorded as to instantaneous voltage and time of
occurrence. The information was transferred to a computer which determined
$(V\text{-}V_{\min})$ for each point in a selected time interval and ordered the results in
ascending sequence. The log and log-log functions were computed and a best
straight line was fitted.

11.7.4 Test programme

The method and procedures described above were implemented by Glinkowski
et al. [4] to compare two contact materials, Cu/Bi and Cu/Cr, in the context of
capacitance switching. A number of interrupters, made by different manufacturers, were employed and some thousands of breakdowns were recorded.
Breakdown conditions did not reconstruct field conditions during capacitance
switching, nevertheless, it was felt that the method provided a good comparative
test for this function.

11.7.5   Discussion of results

Figures 11.9(*a*) and 11.9(*b*) show a comparison between the principal Cu/Cr and
Cu/Bi interrupters for essentially identical conditions. There are sufficient data
(2774 breakdowns) to validate the characteristics. Two features are clear: the
Cu/Cr switch tends to recover more quickly, and reignitions cease in the Cu/Cr
interrupter soon after 2.5 ms whereas they persist in the Cu/Bi device beyond
6 ms. If a reignition occurs in the field during a capacitance switching operation,

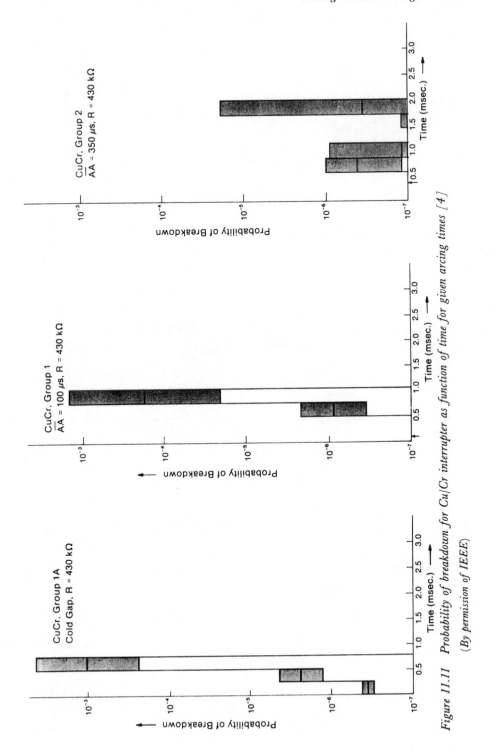

Figure 11.11   Probability of breakdown for Cu/Cr interrupter as function of time for given arcing times [4]

(By permission of IEEE)

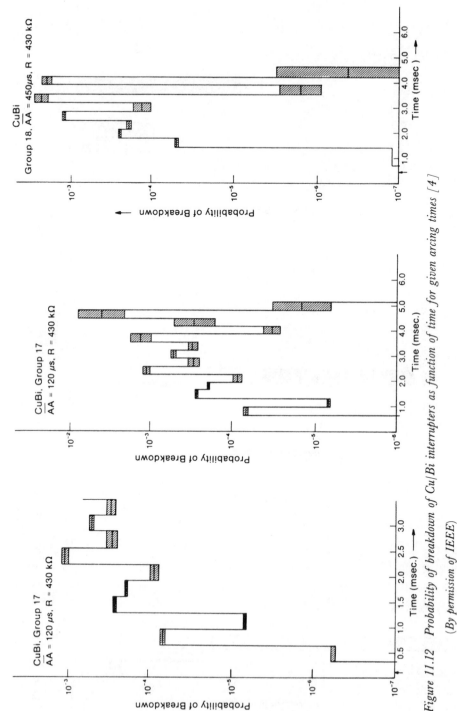

Figure 11.12   Probability of breakdown of Cu/Bi interrupters as function of time for given arcing times [4]

(By permission of IEEE)

it does *not* mean that a serious overvoltage will be generated automatically. For example, if the reignition occurs soon after current zero there will be relatively little voltage across the contacts so the subsequent transient excursion will be limited. Higher voltage transients will occur as the instant of reignition is delayed, being greatest after one half cycle. Such events can lead to voltage escalation as was shown in Section 10.6.3.

The disparity in performance between the two contact materials is reinforced by Figure 11.10 which shows a comparison between Cu/Cr and Cu/Bi for tests with short arcing times. Envelopes are drawn for the upper and lower limits of breakdowns and the average is indicated by the intermediate dots. The information just discussed was qualitative. Now turn to quantitative data derived from statistical analysis. Great accuracy in absolute terms is not claimed for the prediction of breakdown probability. But the figures give a reasonable guide and are certainly very useful for comparing performance of different interrupters.

By way of calibration, use the earlier illustration: a breakdown probability of 10^{-4} means that if there are 50 installations, and if they are switched twice per day, reignitions are likely to occur three or four times a year. It follows that a probability of 10^{-7} would result in three or four reignitions every thousand years in the same circumstances. Thus, a probability of breakdown of 10^{-7} is surely a minimal risk, particularly when, as noted, every reignition does not lead to overvoltage.

Figures 11.11(*a*), (*b*) and (*c*) pertain to the Cu/Cr interrupter. They show how the probability of a breakdown varies as a function of time for given arcing times (the average arcing time, designated $\overline{AA}$ in the caption, is 0, 100 and 350 μs for Figures 11.11(*a*), (*b*) and (*c*), respectively). The numbers have been calculated for a normal 13.8 kV system with ungrounded capacitor banks for which the peak TRV of the first phases to interrupt is

$$2.5 \times \frac{13.8\sqrt{2}}{\sqrt{3}} = 28.17\,\text{kV}$$

The statistics say with 90% confidence, that the breakdown probability is within the shaded areas. The most likely probability is indicated by the line within the box. Beyond 2 ms (or less for Figures 11.11(*a*) and (*b*)) the probability of breakdown is too small to consider, or there is insufficient data to confidently compute a value. This accords with Figure 11.9(*a*).

A set of data for very similar conditions, but for the Cu Bi interrupter is presented in Figures 11.12(*a*), (*b*) and (*c*). Figures 11.12(*a*) and 11.12(*b*) apply to the same group of data. The time scale has been extended in Figure 11.12(*b*) so as to include reignitions at longer times.

These two sets of Figures demonstrate a clear difference between the Cu/Cr and Cu/Bi devices. Three points are worthy of note:

- the breakdown probability is consistently higher for the Cu/Bi interrupter compared with the Cu/Cr interrupter (note - the probability scale for Figure 11.12(*b*) has been shifted by a decade)
- breakdowns persist much longer for the Cu/Bi switch than the Cu/Cr switch
- the confidence band for the Cu/Bi data is very narrow.

11.8   References

1 'Guide to the interpretation of IEC publication 56: 4th edn., 1987 high-voltage alternating current circuit-breakers', *Short Circuit Testing Liaison (STL)* 1988, Rugby CV21 3DW, UK

2 'Alternating current high-voltage power interrupters - safety requirements for x-radiation limits', *ANSI* C37.90, 1989

3 GLINKOWSKI, M., GREENWOOD, A., HILL, J., MAURO, R., and VARNECKAS, V.: 'Comparative switching with vacuum circuit breakers - a comparative evaluation', *IEEE Trans.*, 1991, **PWRD–6**, pp. 1088–1095

4 ROGUSKI, A.T.: ' Experimental investigation of dielectric recovery strength between the separating contacts of vacuum circuit brekers', *IEEE Trans.*, 1989, **PD–4**, pp. 1063–1069

5 MURAI, Y., TOYA, H., and NITTA, T.: 'Statistical properties of the breakdown of vacuum circuit breakers and its influence on the surge generation in capacitance and reactive current interruption', *IEEE Trans.*, 1979, **PAS–98**, pp. 232–238

6 TOYA, H., UENO, N., OKADA, T., and MURAI, Y.: 'Statistical property of breakdown between metal electrodes in vacuum', *IEEE Trans.*, 1981, **PAS–100**, pp. 1932–1939

7 OSMOCROVIC, P., and DJOGO, G.: 'Applicability of simple expressions for the law of breakdown probability increase to electrical breakdown in vacuum', Proceedings of 13th international symposium on Discharges and electrical insulation in vacuum, **1**, 1988, pp. 109–111

8 DJOGO, G., and OSMOCROVIC, P.: 'Statistical properties of electrical breakdown in vacuum', Proceedings of 13th international symposium on Discharges and electrical insulation in vacuum, **1**, 1988, pp. 112–114

Chapter 12

Maintenance of vacuum switchgear

12.1 General comments

This is a short chapter for the good reason that, in truth, modern vacuum switchgear requires very little maintenance. There has been quite a change of attitude regarding maintenance in the last ten years. In 1982 manufacturers were recommending that maintenance be performed '... at least every 1,000 to 2,000 operations, or once a year, whichever comes first'. In 1992, we find a ten-year preventive maintenance schedule being advocated. This change is based on experience during the decade, but in addition it is undoubtedly attributable to improvements in equipment design. Conditions of service (number of operations, kind of environment) remain a factor, but design improvements have made them a less sensitive factor.

It is a matter of record that the incidence of breaker failure tends to increase following maintenance. This suggests that maintenance procedures are sometimes not followed correctly, that when components are disassembled they are not always reassembled in the proper manner. Such 'hazards of maintenance' are less likely to occur where maintenance is less frequent. Of course, this does not imply that no maintenance is the best maintenance of all.

The procedures to be outlined in the ensuing sections apply specifically to a medium-voltage vacuum circuit breaker. However, they are typical of what would be recommended for reclosers and other types of vacuum switchgear.

12.2 Safety

Safety must always be a primary consideration when maintenance of any kind is being performed. It is essential that the breaker be resting securely clamped either on the extended rails outside the switchgear housing or on a transport dolly, and that the control circuits are de-energised, before starting maintenance procedures. Parenthetically, disconnects on either side of a vacuum switch must be open and grounds must be applied to the switch itself, before the equipment is approached for maintenance. Likewise, a vacuum contactor must be disconnected from its power supply before making any inspections or adjustments. Work must never be carried out on a closed breaker/switch/contactor, nor one with its closing springs charged.

12.3 Electrical maintenance

12.3.1 Interrupters

The vacuum interrupter itself is essentially maintenance free. Contact care features prominently in other switching technologies - contacts require cleaning, they wear and must be replaced - but the vacuum interrupter is a sealed unit, the contacts are inaccessible, they are in there for life. Moreover, life is very long in almost all applications; it is very rare that an interrupter must be replaced because it is worn out. Determining the wear that an interrupter has sustained is a very straightforward procedure, since wear or erosion indicators are a standard feature (see Section 11.5, Figure 11.3). Further notes on this matter will be found in Section 12.4.2. On those occasions when it is necessary to replace an interrupter, the task is not difficult or time consuming if the manufacturers instructions are followed. Care should be exercised not to twist the bellows.

Nor is the ambient affected by interrupter operations, indeed, the vacuum often improves with arcing, as noted in Section 1.2. With oil circuit breakers, the oil must be filtered after serious arcing to remove undesirable arc products, and then tested for dielectric strength before being returned to the breaker. Likewise, the gas must be removed from SF_6 breakers periodically. These operations not only take time but require competent personnel and specialised equipment to carry them out.

The only distinctive or unique feature of the vacuum interrupter is its vacuum integrity. At times of major maintenance or if there is any reason to believe that the vacuum may be impaired, a check can be made by the procedure described in Section 11.2

12.3.2   Insulation checks

Concern here is two-fold, first with the integrity of the primary electrical insulation of the breaker as a whole, i.e. the insulation from live parts to ground and from one terminal of the device to another when the interrupter contacts are open, and secondly, with the insulation of the control circuits.

Primary insulation maintenance principally consists of keeping the insulation surfaces clean. This can be done by wiping off all insulating surfaces with a dry lint-free cloth or dry paper towel. Where there is any tightly adhering dirt that will not come off by wiping, it can be removed with a mild solvent or distilled water, making sure that the surfaces are dry before placing the equipment back in service. Detergents should not be used for this purpose as they leave an electrically conducting residue when they dry. This treatment will be required more frequently in certain industrial environments - coal mines, chemical plants, etc. - where pollution is high. While carrying out these procedures, buses and connections should be inspected carefully for evidence of overheating or weakening of the insulation.

The integrity of primary insulation may be checked by an AC high-potential test. The test voltage depends on the maximum rated voltage of the breaker; the values in Table 12.1 are typical. The test is conducted with the breaker closed. A high potential lead is connected between the voltage source and one pole of the device under test; the remaining poles and the equipment frame are grounded. Voltage is increased steadily up to the test value where it is maintained for one minute. The procedure is repeated for the remaining poles. Successful withstand indicates satisfactory insulation strength of the primary circuit.

If a DC high-potential test set is used, rather than an AC source, the peak voltage must not exceed the peak of the corresponding AC test voltage.

Table 12.1   Test voltage for interrupters

Equipment voltage (kV RMS)	4.16	7.2	13.8
Test voltage (kV RMS)	14	27	27

Voltage transformers and control transformers, surge arresters and surge suppressors (where fitted) must be disconnected during high-voltage testing.

To check the secondary circuit insulation, first remove the motor leads. Then connect together all points of the secondary disconnect pins with a piece of hipot wire and attach this in turn to high potential lead of the test set. With the breaker frame grounded, the test set voltage should be increased gradually from zero to 1125 V RMS, where it should be held for one minute. Successful withstand indicates satisfactory insulation strength of the secondary control circuit. Remove the hipot wire following the test and reconnect the motor leads.

12.3.3 Primary circuit resistance check

The DC resistance of the primary circuit may be measured as follows: close the breaker, pass at least 100A DC through the interrupter. With a low-resistance instrument, measure resistance across the studs on the breaker side of the disconnects of each pole. The resistance should not exceed 60, 40 and 20 $\mu\Omega$ for the 1200, 2000 and 3000 A breakers, respectively.

12.4   Mechanical maintenance

12.4.1   Visual inspection

Before making any adjustments it is prudent to make a careful visual inspection of the mechanism for any loose parts such as bolts, nuts, pins, rings, etc. Check excessive wear or damage to the breaker components. Operate the breaker several times manually and electrically to make sure the operation is crisp and without any sluggishness.

12.4.2   Wipe and gap adjustments

Wipe is the additional compression of the preloaded wipe spring (Section 6.2.3, Figure 6.6) which is used to apply force to the closed vacuum interrupter contacts. Proper adjustment of the wipe spring is necessary to assure that the vacuum interrupter contacts will remain closed against the forces that tend to open them due to fault currents and to supply the propelling energy required to attain the correct opening speed needed for a clean interruption of the current.

Gap is the distance between the two vacuum interrupter contacts when the breaker is open. Correct adjustment of the gap assures that the minimum required distance for current interruption is achieved and that the distance is not so great that damage to the vacuum interrupter occurs.

Wipe and gap are related in such a way that decreasing the wipe increases the gap and increasing the wipe decreases the gap. Therefore, these two adjustments must be co-ordinated to bring both to within the required settings

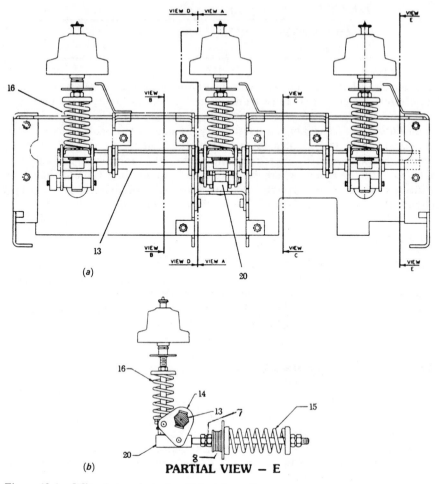

Figure 12.1   Wipe (a) and opening (b) spring adjustments

(Courtesy of General Electric)

simultaneously. The details of how this is done depend on the manufacturer's design, but the principle is similar from one design to another. It is illustrated by the arrangement shown in Figure 12.1. Observe two springs in Figure 12.1(*b*), the wipe spring (16) in the vertical position and the opening spring (15) in the horizontal position. The action of opening and closing the contacts is effected by rotating the hexagonal operating shaft which can be seen in section, crosshatched, as item (13). This rotates the bell crank (14) which in turn moves the operating rod (5) up or down. Rotating the bell crank clockwise closes the contacts. However, once they have mated, the bell crank can continue to rotate through a further small angle and compress the wipe spring an amount d, thereby applying a force xd between the contacts, x being the wipe spring gradient.

The distance d is evident from an indicator. It can be increased or decreased by loosening the wipe lock nut (6) and rotating the operating rod which virtually changes its length; screwing the operating rod towards the interrupter increases the wipe. The threading arrangement is, of course, such that the interrupter itself is not rotated. After making a wipe adjustment, the wipe lock nut must be retightened.

When a trip signal is given the bell crank proceeds to rotate counter-clockwise, being driven by the combined efforts of the wipe spring and the opening spring. The contacts do not separate until the wipe spring has extended its permitted distance d. Figure 12.1 shows the breaker pole in the open position with the gap adjusting nut (7) against the stack of washers (8). In the closed position there is a space between the gap adjusting nut and the washers, this space is the sum of the wipe d, and the gap length, due allowance being made for differences in radii of turning circles of the two pins on the bell crank. It is therefore evident that the gap can be changed by advancing the nut 7 to the right (to shorten the gap) or to the left (to lengthen the gap). Nut 7 should be locked with its lock nut after making any adjustment.

If a breaker executes a large number of short-circuit current interruptions, the contacts will be eroded, material will be boiled off and deposited on the shields. The moving contact must therefore move further into the interrupter to mate with the fixed contact. Evidence that this has occurred will be seen when the erosion disc (8) which is attached to operating rod, no longer lines up with the erosion indicator (9) (attached to the breaker frame). The consequence of erosion is to reduce the wipe and increase the gap.

The manufacturer's instructions will state the tolerable limits for wipe and gap. When these are approached adjustments should be made. The wipe is restored first, by the procedure just outlined. The gap plus wipe is then adjusted by repositioning the gap adjusting nut (7).

The manufacturer will also recommend a maximum value for erosion; this is typically 0.125 in, or 3.2 mm.

12.4.3 Timing

Timing may be checked by monitoring the control circuit voltage and by using a low voltage signal through the vacuum interrupter contacts to indicate the closed or open position. Typical time ranges vary with coil voltages but nominal values are:

Initiation of trip signal to contact parting	
5 cycle (60 Hz) breaker	3.5 to 5.0 ms
3 cycle (60 Hz) breaker	2.5 to 3.0 ms
Initiation of close signal to contact closing	
standard breaker	60 to 100 ms
fast bus transfer breaker	62 ms max
Instantaneous reclose time*	128 to 221 ms.

* Time from application of trip signal and close signal until breaker opens and recloses

12.4.4   Lubrication

All parts that require lubrication are lubricated during assembly, molybdenum disulphide grease is often used. This lubricant may disperse or degrade over a period of time, it should therefore be replaced when scheduled maintenance is carried out. The following locations should be lubricated with light machine oil, sparingly applied:

(i)   wire link ends
(ii)   spring charged/discharged indicator cam surface
(iii)   trip latch surface/trip pin mating surfaces
(iv)   spring release latch/spring release pin mating surfaces
(v)   closing and opening springs crank pins
(vi)   bell crank pins in pole unit area

After lubrication the breaker should be operated making sure that opening and closing operations are crisp and snappy.

Roller bearings are often used on the pole shaft, the cam shaft, the main link and the motor eccentric. These bearings are packed at the factory with a top-grade slow-oxidising grease which normally should be effective for many years. They should not be disturbed unless there is definite evidence of sluggishness or dirt, or unless the parts are dismantled for some reason.

If it becomes necessary to disassemble the mechanism, the bearings and related parts should be thoroughly cleaned of old grease in a good grease solvent. Do not use carbon tetrachloride. They should then be washed in light machine oil until the cleaner is removed. After the oil has been drawn off, the bearings should be packed with the manufacturer's recommended lubricant.

12.5   Maintenance of cubicle and auxiliaries

The following items should receive attention:

(i)   *Racking mechanism:* clean and lubricate the jack screws and gears.
(ii)   *Primary disconnects:* check contacts for signs of abnormal wear or overheating. Discolouration of the silvered surfaces is not ordinarily harmful unless atmospheric conditions cause deposits such as sulfides on the contacts. If necessary the deposits can be removed with a good grade of silver polish. Sandpaper, steel wool or abrasive cleaners should never be used on silver-plated parts. Before replacing the breaker, apply a thin coat of approved contact lubricant to the breaker studs.
(iii)   *Anchor bolts:* see that all anchor bolts and other bolts in the structure are tight.
(iv)   *Heaters:* if the switchgear is equipped with heaters, check to see that all heaters are energised and operating. Inspect individual heater elements and replace any that have failed.
(v)   *Filters:* All louvered exterior openings in outdoor equipment are furnished with air filters. The foam filter elements should be removed, washed in soapy water, dried and then reassembled annually. Elements should be inspected before reassembly if any signs of deterioration are evident.

Author index

Subject index

Accommodation coefficient 63, 134
Advanced pole opening 219
Air insulated
  distribution switchgear 184
  equipment 182
Air magnetic circuit breaker 10
Alumina 126
Anode spot 27, 41, 46
Antenna switch 5
Antimony 110
Arc furnace switching 132
Arc mode
  constricted 10, 25, 27, 76
  diffuse 25
Arc voltage 3, 15, 29, 61, 210
Asymmetrical fault current 214
Atom-ion balance 85

Backfeed (from motors) 212
Bakeout and bakeout ovens 10, 202
Band theory of metals 35
Bellows 3, 108, 121
  annealing of 122
  antibuckling 124
  fatigue life 121
  shield 125
Bimetal contact 17
Blow off force
  see popping force
Brazing 200
  "one shot" braze 203

Cable switching 227
Cam-follower mechanism 152
Capacitance switching 6, 217, 262
  bank-to-bank 222
  cable 227
  line dropping 280
Cathode spot 25
  model for 34
Chemical etching 200
Chrome/copper 17, 110, 112, 113, 197, 262
  grain size 198
Circuit resistance check 268
Clean room 201, 205
Cleaning 199
Close and latch rating 209
"Clump" 57
Conditioning 3, 56

Constricted arc 25, 27
  interrupting 76
Contacts
  bimetal 17
  bounce and rebound 146
  button-type 114
  contrate 17, 18, 76, 114
  dynamics 146
  erosion of 27, 76, 210, 224
  geometry 114
  grain size 198
  hard 76
  manufacture of 297
  material 109
  popping 79
  resistance 120
  soft 76
  spiral 10, 76, 114
Contrate contact 17, 18, 86, 114
Conversion kits 184
Copper/bismuth 110, 199, 262
Copper/chrome 17, 110, 112, 113, 197, 262
Copper/lead 113
Copper/silver/palladium 201
Copper/tungsten 113
Corona 53
Current chopping 9, 10, 25, 81, 216

D'Alambert's principle 145
Deconditioning 56
Degassing 201
Degreasing 199
Degree of vacuum 109
Degrees of freedom 50
De-ionized water 199
Dielectric breakdown 200
Diffuse arc 25
  interrupting 62
  maintaining 77
Direct current interruption 96, 235
Directional cooling 111
Displacement current 74, 94
Dry-type transformer 217

Electromagnetic force 178, 209
Electromagnetic pickup 224
Electron avalanche 52
Electron emission 35, 56
  field 35, 56